Why Has Nobody
Told Me This Before?

Why Has Nobody Told Me This Before?

DR JULIE SMITH

HarperOne
An Imprint of HarperCollins*Publishers*

HarperCollins books may be purchased for educational, business, or sales promotional use. For information, please email the Special Markets Department at SPsales@harpercollins.com.

Originally published as *Why Has Nobody Told Me This Before?* in Great Britain in 2022 by Michael Joseph.

FIRST HARPERONE HARDCOVER PUBLISHED IN 2022

Library of Congress Cataloging-in-Publication Data has been applied for.

ISBN 978-0-06-322793-4

23 24 25 26 27 LBC 22 21 20 19 18

For Matthew.

If mine is the ink then yours is the paper. Like all our adventures we got here together.

Contents

Introduction   1

1: On Dark Places

1: Understanding low mood   9
2: Mood pitfalls to watch out for   20
3: Things that help   34
4: How to turn bad days into better days   44
5: How to get the basics right   53

2: On Motivation

6: Understanding motivation   67
7: How to nurture that motivation feeling   72
8: How do you make yourself do something
      when you don't feel like it?   82
9: Big life changes. Where do I start?   94

3: On Emotional Pain

10: Make it all go away!   99
11: What to do with emotions   105
12: How to harness the power of your words   114
13: How to support someone   120

Contents

4: On Grief

14: Understanding grief   127
15: The stages of grief   132
16: The tasks of mourning   138
17: The pillars of strength   146

5: On Self-doubt

18: Dealing with criticism and disapproval   151
19: The key to building confidence   162
20: You are not your mistakes   173
21: Being enough   179

6: On Fear

22: Make anxiety disappear!   189
23: Things we do that make anxiety worse   195
24: How to calm anxiety right now   200
25: What to do with anxious thoughts   205
26: Fear of the inevitable   223

7: On Stress

27: Is stress different from anxiety?   233
28: Why reducing stress is not the only answer   239
29: When good stress goes bad   244
30: Making stress work for you   251
31: Coping when it counts   264

8: On a Meaningful Life

32: The problem with 'I just want to be happy' 277
33: Working out what matters 284
34: How to create a life with meaning 293
35: Relationships 297
36: When to seek help 317

References 321
Resources 333
Acknowledgements 335
Index 339
Spare tools 347

Introduction

I'm sitting in my therapy room across from a young woman. She is relaxed in the chair, her arms open and loosely moving as she speaks to me. A transformation from the tension and nerves of her first session. We have only had a dozen appointments. She looks into my eyes and starts to nod and smile as she says, 'You know what? I know it's going to be hard, but I know I can do it.'

My eyes sting and I swallow. The smile sweeps across every muscle in my face. She has felt the shift and, now, so have I. She came into this room, some time ago, fearful of the world and everything she had to face. Pervasive self-doubt led her to feel dread for every new change and challenge. She left therapy that day with her head held a little higher. Not because of me. I have no magical ability to heal anyone or change their life. She had not needed years of therapy that unravelled her childhood. In this situation, as in many others, the major part of my role was as an educator. I passed on insights about what the science says and what has worked for others. Once she understood and started using the concepts and skills, a transformation began. She felt hope for the future. She started to believe in her own strength. She started dealing with difficult situations in healthy new ways. Each time she did, confidence in her ability to cope grew a bit more.

Introduction

As we revisited the things she needed to remember in order to face the week ahead, she nodded, looked at me and asked, 'Why has nobody told me this before?'

Those words stayed with me, ringing in my head. She was not the first or the last person to say them. The same scenario repeated itself over and over. Individuals were coming along to therapy believing that their strong painful emotions were the result of a fault in their brain or personality. They did not believe they had any power to influence them. While longer-term, more in-depth therapy is appropriate for some people, there were so many who simply needed some education about how their mind and body work and how they could manage their mental health day-to-day.

I knew the catalyst was not me, it was the knowledge they were being introduced to. But people should not have to pay to come and see someone like me just to get access to that education about how their mind works. Sure, the information is out there. But in a sea of misinformation, you have to know what you are looking for.

I started campaigning into my poor husband's ear about how things should be different. 'OK, go for it,' he said. 'Put some videos on YouTube or something.'

So we did. Together we started making videos talking about mental health. As it turned out, I was not the only one who wanted to talk about this stuff. Before I knew it, I was making almost daily videos for millions of followers across social media. But the plat-forms where I could reach the most people seemed to be those with short-form videos. This means I have a large collection of videos with no longer than 60 seconds to get my point across.

While I have been able to catch people's attention, share some insights and get them talking about mental health, I still want to go one step further. When you make a 60-second video there is so much that you have to leave out. So much detail that gets missed. So, here it is. The detail. The ins and outs of how I might explain some of these concepts in a therapy session and some simple guidance on how to use them, step by step.

The tools in this book are mostly taught in therapy, but they are not therapy skills. They are life skills. Tools that can help every single one of us to navigate through difficult times and to flourish.

In this book, I will break down the things I have learned as a psychologist and gather together all of the most valuable knowledge, wisdom and practical techniques I have come across that have changed my life and those of the people I have worked with. This is the place to get clarity on emotional experience and a clear idea of what to do about it.

When we understand a little about how our minds work and we have some guideposts on how to deal with our emotions in a healthy way, we not only build resilience, but we can thrive and, over time, find a sense of growth.

Before leaving their first therapy session, many people want some sort of tool they can take home and start using to ease their distress. For this reason, this book is not about delving into your childhood and working out how or why you came to struggle. There are other great books for that. But, in therapy, before we can expect anyone to work on healing any past traumas, we must ensure they have the tools in place to build resilience and the ability to tolerate distressing emotions safely. There is such power in understanding the many ways you can influence how you feel and nurture good mental health.

This book is all about doing just that.

This book is not therapy, in the same way that a book about how to maximize your physical health is not medicine. It is a toolbox filled to the brim with different tools for different jobs. You cannot master how to use them all at the same time, so you don't need to try. Pick the section that fits with the challenges you face right now, and spend time applying those ideas. Every skill takes time to become effective, so give it a chance and plenty of repetition before you discard any of the tools. You cannot build a house with just one tool. Each task requires something slightly different. And however skilled you get at using those tools, some challenges are just much harder than others.

To me, working on maximizing our mental health is no different to working on our physical health. If you put health on a number scale with zero as neutral – not unwell but not thriving – a number below zero would indicate a health problem and any number above zero would indicate good health. In the last few decades it has become acceptable and even fashionable to work on maximizing your physical health through nutrition and exercise. Only more recently has it become acceptable to openly and visibly work on your mental health. This means you don't need to wait until you're struggling before you pick up this book, because it is OK to build upon your mental health and resilience, even if you are not unwell or struggling right now. When you feed your body with good nutrition and build up stamina and strength with regular exercise, you know that your body is more able to fight infection and heal when faced with injury. It's just the same with mental health. The more work we do on building self-awareness and resilience when all is well, the better able we are to face life's challenges when they come our way.

If you pick a skill from this book and find it useful, in hard times don't stop practising that skill when everything starts to improve. Even when you are feeling good and don't think you need it, these skills are nutrition for your mind. It's like paying a mortgage rather than rent. You are investing in your future health.

The things included in this book have a research evidence base. But I do not rest on that alone. I also know they can help because I have seen them help, time and time again, for real people. There is hope. With some guidance and self-awareness, struggle can build strength.

When you start to share things on social media or you write a self-help book, lots of people get the impression that you have it all sorted. I have seen a lot of authors in the self-help industry perpetuate this idea. They feel they have to look as if the things life throws at them leave no dents or scars. They suggest that their book contains the answers – all the answers you will ever need in life. Let me demystify that one right now.

I am a psychologist. That means I have read a lot of the

research that has been produced on this subject and I have been trained to use it to help guide other people in their quest towards positive change. I am also a human. The tools I have acquired do not stop life throwing stuff at you. They help you to navigate, swerve, take a hit and get back up. They don't stop you getting lost along the way. They help you to notice when you have lost your way and bravely turn on your heel and head back towards a life that feels meaningful and purposeful to you. This book is not the key to a problem-free life. It is a great bunch of tools that helps me and many others find our way through.

The journey so far . . .

I am not a guru who has all the answers to the universe. This book is part journal, part guide. In some ways I have always been on a personal quest to discover how it all pieces together. So this book is me making use of all those hours spent reading, writing and speaking with real humans in therapy to understand a bit more about being human and what helps us while we are here. This is only the journey so far. I continue to learn and be amazed by people I meet. Scientists keep asking better questions and discovering better answers. So here is my collection of the most important things I have learned *so far* that have helped both me and the people I work with in therapy to find our way through human struggle.

So this book is not necessarily going to ensure that you live the rest of your days with a smile on your face. It will let you know which tools you can use to make sure that when you do smile, it is because you genuinely feel something. It will describe the tools you need to keep re-evaluating and finding your direction, returning to healthier habits and self-awareness.

Tools might look great in the box. But they only help when you get them out and start practising how to use them. Each tool takes regular practice. If you miss the nail with the hammer this time, come back later and try again. As a fellow human being, I too continue to do this, and I have only included techniques and skills that I have tried and found useful both for myself and for the individuals I have worked with. This book is a resource for me as much as it is for you. I will keep returning to it time and time again whenever I feel I need to. My wish is that you will do the same and that it can be a toolbox for life.

1

On Dark Places

CHAPTER 1

Understanding low mood

Everyone has low days.

Everyone.

But we all differ in how frequent the low days are and how severe the low mood.

Something that I have come to realize over the years of working as a psychologist is how much people struggle with low mood and never tell a soul. Their friends and family would never know. They mask it, push it away and focus on meeting expectations. Sometimes people arrive at therapy after years of doing that.

They feel like they're getting something wrong. They compare themselves to the people who appear to have it all together all of the time. The ones who are always smiling and apparently full of energy.

They buy into the idea that some people are just like that and happiness is some sort of personality type. You either have it or you don't.

If we see low mood as purely a fault in the brain, we don't believe we can change it, so instead we get to work on hiding it. We go about the day, doing all the right things, smiling at all the right people, yet all the time feeling a bit empty and dragged down by that low mood, not enjoying things in the way we are told we should.

Take a moment to notice your body temperature. You might feel perfectly comfortable, or you may be too hot or too cold. While changes in how hot or cold you feel could be a sign of infection and illness, it could just as easily be a signal of things around you. Maybe you forgot your jacket, which is normally enough to protect you from the cold. Perhaps the sky has clouded over and it has started to rain. Maybe you are hungry or dehydrated. When you run for the bus you notice you warm up. Our body temperature is affected by our environment, both internal and external, and we also have the power to influence it ourselves. Mood is much the same. When we experience low mood, it may have been influenced by several factors from our internal and external world, but when we understand what those influences are, we can use that knowledge to shift it in the direction we want it to go. Sometimes the answer is to grab an extra layer and run for the bus. Sometimes it's something else.

Something that the science has been confirming to us, and something people often learn in therapy, is that we have more power to influence our emotions than we thought.

This means we get to start working on our own wellbeing and taking our emotional health into our own hands. It reminds us that our mood is not fixed and it does not define who we are; it is a sensation we experience.

This doesn't mean we can eradicate low mood or depression. Life still presents us with hardship, pain and loss and that will always be reflected in our mental and physical health. Instead, it means we can build up a toolbox with things that help. The more we practise using those tools, the more skilled we get at using them. So when life throws us problems that hammer our mood into the ground we have something to turn to.

The concepts and skills covered are for us all. Research shows them to be helpful for those with depression, but they are not a controlled drug that you need a prescription for. They are life skills. Tools that we can all use as we go through life facing fluctuations in mood, big and small. For anyone who experiences severe and enduring mental illness it is always optimal to learn new skills with the support of a professional.

How feelings get created

Sleep is bliss. Then my alarm offends my ears. It's too loud and I hate that tune. It sends a shockwave through my body that I am not ready for. I press snooze and lie back down. My head is aching and I feel irritated. I press snooze again. If we don't get up soon the kids will be late for school. I need to get ready for my meeting. I close my eyes and see the to-do list lying on my desk in the office. Dread. Irritation. Exhaustion. I don't want to do today.

Is this low mood? Did it come from my brain? How did I wake up like this? Let's trace back. Last night I stayed up late working. By the time I got into bed I was too tired to go back downstairs to grab a glass of water. Then my baby woke up twice in the night. I haven't slept enough and I'm dehydrated. The loud alarm woke me from a deep sleep, sending stress hormones shooting through my body as I woke up. My heart started pounding and that felt something like stress.

Each of these signals sends information to my brain. We are not OK. So my brain goes on a hunt for reasons why. It searches, it finds. So my physical discomfort, brought about by lack of sleep and dehydration, helped to create low mood.

Not all low mood is unidentified dehydration, but when dealing with mood it is essential to remember that it's not all in your head. It's also in your body state, your relationships, your past and present, your living conditions and lifestyle. It's in everything you do and don't do, in your diet and your thoughts, your movements and memories. How you feel is not simply a product of your brain.

Your brain is constantly working to make sense of what is going on. But it only has a certain number of clues to work from. It takes information from your body (e.g. heart rate, breathing, blood pressure, hormones). It takes information from each of your senses – what you can see, hear, touch, taste and smell. It takes information from your actions and thoughts. It pieces all these clues together with memories of when you have felt similar in the past and makes a suggestion, a best guess about what is happening and what you do about it. That guess can sometimes be felt as an emotion or a mood. The meaning we make of that emotion and how we respond to it, in turn, sends information back to the body and the mind about what to do next (Feldman Barrett, 2017). So when it comes to changing your mood, the ingredients that go in will determine what comes out.

The two-way road

Lots of self-help books tell us to get our mindset right. They tell us, 'What you think will change how you feel.' But they often miss something crucial. It doesn't end there. The relationship works both ways. The way you feel also influences the types of thoughts that can pop into your head, making you more vulnerable to experiencing thoughts that are negative and self-critical. Even when we know our thought patterns aren't helping, it is so incredibly hard to think differently when we feel down, and even harder to follow the rule of 'only positive thoughts' that is often suggested on social media. The mere presence of those negative thoughts does not mean that they came first and caused the low mood. So thinking differently may not be the only answer.

How we think is not the whole picture. Everything we do and don't do influences our mood too. When you feel down, all you want to do is hide away. You don't feel like doing any of the things you normally enjoy, and so you don't. But disengaging from those things for too long makes you feel even worse. The loop also occurs with our physical state. Let's say you have been too busy to exercise for a few weeks. You feel tired and low in mood, so exercising is the last thing you want to do. The longer you avoid the exercise, the more you feel lethargic and low on energy. When you are low on energy, the chance of exercising goes down, along with your mood. Low mood gives you the urge to do the things that make mood worse.

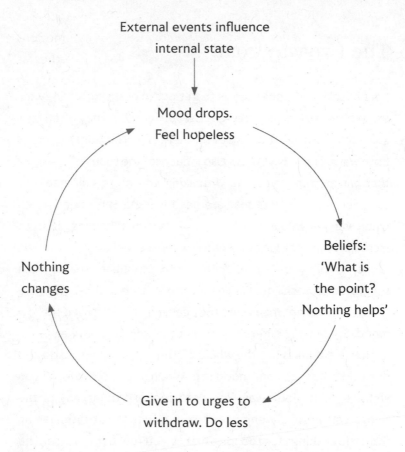

Figure 1: The downward spiral of low mood. How a few days of low mood can spiral into depression. Breaking the cycle is easier to do if we recognize it early and act on it. Adapted from Gilbert (1997).

So we get into these vicious cycles easily because all the different aspects of our experience are impacting each other. But while this shows us how we can get stuck in a rut, it also shows us the way out.

All these things are interacting to create our experience. But we don't experience our thoughts, bodily sensations, emotions and actions all separately. We experience them together as one. Like wicker strands woven together, it's hard to notice each one individually. We just experience the basket as a whole. That is why we need to get practised at breaking it down. When we do that we can more easily see what changes we could make. Figure 2, on the next page, shows a simple way to break down your experience.

When we break things down in this way, we can start to recognize not only what we do that keeps us stuck but also what we do that helps.

Most people come to therapy knowing that they want to feel different. They have some unpleasant (sometimes excruciating) feelings they don't want to have any more and are missing some of the more enriching emotions (such as joy and excitement) that they would like to feel more of. We can't just press a button and produce our desired set of emotions for the day. But we do know that how we feel is closely entwined with the state of our body, the thoughts we spend time with and our actions. Those other parts of our experience are the ones that we can influence and change. The constant feedback between the brain, the body and our environment means that we can use those to influence how we feel.

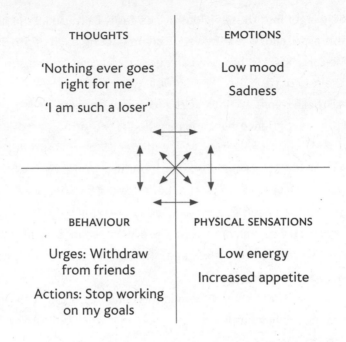

THOUGHTS

'Nothing ever goes
right for me'

'I am such a loser'

EMOTIONS

Low mood

Sadness

BEHAVIOUR

Urges: Withdraw
from friends

Actions: Stop working
on my goals

PHYSICAL SENSATIONS

Low energy

Increased appetite

Figure 2: Spending time with negative thoughts makes it highly likely that I will feel low in mood. But feeling low in mood also makes me more vulnerable to having more negative thoughts. This shows us how we get stuck in cycles of low mood. But it also shows us the way out. Adapted from Greenberger & Padesky (2016).

Where to start

The first step to begin getting a grasp on low mood is to build our awareness of each aspect of the experience. This simply means noticing each one. This awareness starts off with hindsight. We look back on the day and choose moments to look at in detail. Then, with time and practice, that builds our ability to notice them in the moment. This is where we get the opportunity to change things.

In therapy I might ask someone with low mood to notice where they feel it in their body. They might notice that they feel tired and lethargic or lose their appetite. They might also notice that when they feel low they have thoughts like, 'I don't feel like doing anything today. I am so lazy. I'll never be successful. What a loser.' They might have the urge to go back to hide in the bathroom at work for a while and scroll through social media.

Once you get familiar with what is going on inside your own body and mind, you can then expand that awareness to looking at what is going on in your environment and your relationships and the impact that is having on your internal experience and behaviour. Take your time getting to know the details. *When I am feeling this, what am I thinking about? When I am feeling this, what state is my body in? How was I looking after myself in the days or hours leading up to this feeling? Is this an emotion or just physical discomfort from an unmet need?* There are lots of questions. Sometimes the answers will be clear. Other times it will all feel too complex. That is OK. Continuing to explore and write down experiences will help to build up self-awareness about what makes things better and what makes things worse.

 Toolkit: Reflect on what is contributing to your low mood

Use the cross-sectional formulation (see Figure 2, page 16) to practise the skill of picking up on the different aspects of

experiences, both positive and negative. You'll find a blank formulation on page 347 that you can fill in yourself. Take 10
minutes and pick a moment from that day to reflect on. You
may notice that some boxes are easier to fill than the others.

Reflecting on moments after they happen will help to gradually build up the skill of noticing the links between those
aspects of your experience as they happen.

 Try this: You can use these prompts to help you fill in
the formulation. Or you can simply use these as journal
prompts.

- What was happening in the lead up to the moment
  you are reflecting on?
- What was happening just before you noticed the
  new feeling?
- What were your thoughts at the time?
- What were you focusing your attention on?
- What emotions were present?
- Where did you feel that in your body?
- What other physical sensations did you notice?
- What urges appeared for you?
- Did you act on those urges?
- If not, what did you do instead?
- How did your actions influence the emotions?
- How did your actions influence your thoughts and
  beliefs about the situation?

Chapter summary

- Mood fluctuation is normal. Nobody is happy all the time. But we don't have to be at the mercy of it either. There are things we can do that help.
- Feeling down is more likely to reflect unmet needs than a brain malfunction.
- Each moment of our lives can be broken down into the different aspects of our experience.
- Those things all influence each other. It shows us how we get stuck in a downward spiral of low mood or even depression.
- Our emotions are constructed through a number of things we can influence.
- We cannot directly choose our emotions and switch them on but we can use the things we can control to change how we feel.
- Using the cross-sectional formulation (see Figure 2, page 16) helps to increase awareness of what is impacting on our mood and keeping us stuck.

CHAPTER 2

Mood pitfalls to watch out for

The problem with instant relief

Low mood gives us the urge to do things that can make our mood even worse. When we feel discomfort and the threat of low mood, we want to get back to feeling lighter. Our brain already knows from experience what tends to help the quickest. So we feel urges to do whatever will make it all go away as soon as possible. We numb or distract ourselves, and push the feelings away. For some that is via alcohol, drugs or food. For others it is watching hours of TV or scrolling through social media. Each of those things are so inviting because they work – in the short term. They give us that instant distraction and numbing that we crave. That is, until we switch off the TV, close down the app, or sober up, and then the feelings come back. Each time we go round that cycle the feelings come back even more intense.

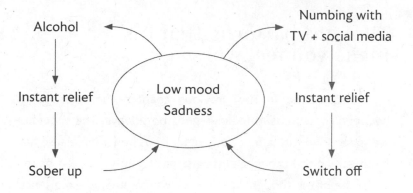

Figure 3: The vicious cycle of instant relief. Adapted from the work of Isabel Clark (2017).

Finding ways to manage low mood involves reflecting on the ways in which we respond to those feelings, having compassion for our human need for relief, while also being honest with ourselves about which of those attempts to cope are making things worse in the longer term. Often the things that work best in the long term are not fast-acting.

 Try this: Use these questions as journal prompts to help you reflect on your current coping strategies for low mood.

- When feeling low, what are your go-to responses?
- Do those responses provide instant relief from the pain and discomfort?
- What effect do they have in the long term?
- What do they cost you? (Not in money, but in time, effort, health, progress.)

Thought patterns that make you feel worse

As we discussed in the previous chapter, the relationship between thoughts and feelings goes both ways. The thoughts we spend time with affect how we feel, but how we feel also has an effect on the thought patterns that come up. Listed below are some of the thought biases that we commonly experience when mood is low. They might sound familiar and that's because thought biases are normal. They happen to everyone to varying degrees. But they are more likely to happen when we experience fluctuations in mood and emotional states. Understanding what they are and starting to notice them when they appear is a big step towards taking some of the power out of them.

Mind reading

Having a grasp on what the people around us are thinking and feeling is crucial for humans. We live in groups and depend on each other, so we all spend much of our lives making guesses about what other people are thinking and feeling. But when we're feeling down, we are more likely to assume that those guesses are true. 'When my friend looked at me funny I just *knew* she hated me.' But on a different day, when I'm not struggling with low mood, I might be more inclined to be a bit more curious about what was going on and possibly even ask her.

You might notice that you feel the need for more reassurance from others when your mood is low. If you don't get that extra reassurance you might automatically assume that they are thinking negatively about you. But that is a bias, and it is quite possible that you are your worst critic.

Overgeneralization

When we are struggling with low mood it only takes one thing to go wrong, and we have that tendency to write off the whole day. You spill some milk in the morning. It goes everywhere. You feel stressed and frustrated as you don't want to be late. Overgeneralization is when we see this one event as a sign that today will be 'one of those days'. Nothing is going your way, it never does. You start asking the universe to give you a break because it sure as hell feels like it's against you today.

When this happens we start to expect more things to go wrong and it's a slippery slope towards hopelessness. Overgeneralizing thoughts particularly like to show up along with the pain of a breakup. One relationship ends and our thoughts start to suggest that this means we will never make a relationship work and could never be happy with anyone else. It is natural to have these thoughts, but left unchecked they will contribute to more pain and low mood.

Egocentric thinking

When times are hard and you're not feeling at your best, this tends to narrow our focus. It becomes more difficult to consider other people's opinions and perspectives, or that they might hold different values. This bias can cause problems in our relationships because it can disrupt how connected we feel to others.

For example, we set ourselves a rule for living, something like, 'I must always be on time for everything.' We then apply that rule to others and feel offended or hurt when they fall short of that. That might make us feel less tolerant of others, disrupt our mood even further and add relationship tensions into the mix. This equates with trying to control the uncontrollable and inevitably sends our low mood spiralling down further.

Emotional reasoning

Just as thoughts are not facts, feelings are not facts either. Emotions are information, but when that information is powerful, intense and loud, as emotions can be, then we are more vulnerable to believing in them as a true reflection of what is going on. *I feel it therefore it must be a fact.* Emotional reasoning is a thought bias that leads us to use what we feel as evidence for something to be true, even when there might be plenty of evidence to suggest otherwise. For example, you walk out of an exam feeling deflated, low in mood and lacking in confidence. Emotional reasoning tells you this means you must have failed. You may have performed OK in the exam, but your brain takes information from how you feel and you're not feeling like a winner right now. The low mood could have been created by the stress followed by exhaustion, but the feeling is influencing how you then interpret your situation.

The mental filter

The thing about the human brain is that, when you believe something, the brain will scan the environment for any signs that the belief is true. Information that challenges our beliefs about ourselves and the world is psychologically threatening. Things suddenly become unpredictable and that doesn't feel safe. So the brain tends to discount it and hold on to whatever fits with previous experience, even if that belief causes distress. So during hard times, when you may be feeling low and believe that you are a failure, your mind will act like a sieve, letting go of all the information that suggests otherwise, and holding on to any indication that you have not lived up to expectation.

Let's say you post a picture on social media and plenty of your followers leave positive comments. But you are not looking for those. You skim past them, searching for any negative ones. If you find any, you might then spend a significant portion of your day thinking it over, feeling hurt, and doubting yourself.

In evolutionary terms, it makes sense that when you feel vulnerable, you keep an extra lookout for signs of threat. But when you are trying to come back from a dark place, the mental filter is something to be aware of.

Musts and shoulds

Beware of those *musts* and *shoulds*! I don't mean the healthy and normal sense of duty we have to our community. I mean the relentless expectations that send us on a downward spiral of unhappiness. *I must be more this*, and *I should feel that*.

The musts and shoulds are heavily tied up with perfectionism. For example, if you feel you must never fail, you are setting yourself up for a rollercoaster of emotions and a struggle with mood when you make a mistake or encounter a setback. We can strive for success and accept failures along the way. But when we set ourselves unrealistic expectations, we become trapped by them. That means we suffer whenever there is any sign that we may not be living up to them.

So watch out for those musts and shoulds. When you are already struggling with mood, expecting yourself to do, be and have everything that you are when you're at your best is not realistic or helpful.

All-or-nothing thinking

Also known as black-and-white thinking, this is another thought bias that can make mood worse if we leave it unchecked. This is when we think in absolutes or extremes. *I am either a success or a complete failure. If I don't look perfect, I'm ugly. If I make a mistake, I should never have bothered.* This polarized thinking style leaves no room for the grey areas that are often closer to reality. The reason this pattern of thought makes everything harder is because it makes us vulnerable to more intense emotional reactions. If failing one exam means you are a failure as a person, then the emotional fallout from that will be more extreme and much harder to pull back from.

When you feel low in mood, you're more likely to think in this polarized way. But it's important to remember that this is not because your brain is getting things wrong or malfunctioning in any way. When we are under stress, all-or-nothing thinking creates a sense of certainty or predictability about the world. What we then miss is the chance to think things through more logically, weighing up the different sides of the argument and coming to a more informed judgement.

Figure 4: Table of thought bias examples.

THOUGHT BIAS	WHAT IS IT?	EXAMPLE
Mind reading	Making assumptions about what others are thinking and feeling.	'She hasn't called in a while because she hates me.'
Overgeneralization	Taking one event and using it to generalize about other things.	'I failed my exam. My future is ruined.'
Egocentric thinking	Assuming that others have the same perspective and values as we have, and judging their behaviour through that lens.	'I would never be late like that. He obviously doesn't care enough about me.'
Emotional reasoning	I feel it, therefore it must be true.	'I feel guilty, therefore I am a bad parent.'
Musts and shoulds	Relentless and unrealistic expectations that set us up to feel like a failure every day.	'I must always look perfect.' 'I should never do any less than my absolute best.'

All-or-nothing thinking	Thinking in absolutes or extremes.	'If I don't get 100 per cent I'm a failure.' 'If I don't look perfect I'm not going out.'

What to do with thought biases

Now you know some of the common thought biases that can make your mood worse, what next? We can't stop those thoughts from arriving, but the power is all in seeing them for what they are (biased) and then managing how we respond to them. If we can acknowledge that each of our thoughts presents just one possible idea among many, then we open ourselves up to the possibility of considering others. This means the original thought has less power over our emotional state.

To be sure we respond to them in the way we want to, first we need to notice the biases when they appear. If we don't step back and see them as a bias, we buy into them as if they present a fair reflection of reality. Then they can feed that low mood and exert their influence over what we do next.

Noticing thought biases sounds obvious, and it is simple. But it's not always easy. When we're in the moment, we don't only experience a thought that we can see clearly. We experience the mess of emotions, physical sensations, images, memories and urges, all at once. We are so used to doing everything on autopilot that stopping to check out the details of the process can take a lot of practice.

Here are some ways you can start to spot thought biases and the impact they have on you.

Getting started

- High emotion states can make it hard to think clearly, so it can be easier to start by reflecting on thought biases after the emotions of the moment have passed. You build your awareness by looking back, but that gradually builds towards awareness in real time.
- Start keeping a journal and choose specific moments to focus on (both positive and negative). Make a distinction between what you were thinking at the time, what emotions you noticed and what physical sensations came with that. Once you have the thoughts written down, go over the list of biases and see if your thoughts might have been biased at the time.
- If you are in the moment and have the chance to write something down, put pen to paper and express your thoughts, feelings and bodily sensations. But as you do that, try to use language that helps you get some distance from those thoughts and feelings. For example, *I am having thoughts that . . .* or *I am noticing these sensations.* This use of language helps you to step back from the thoughts and feelings, to see them as an

experience that is washing over you, rather than an absolute truth.

- If you have someone you trust and confide in, you can share with them the thought biases that you are prone to and they can help you to spot them and call them out. But this requires a very good relationship with someone who is accepting and respectful and supports you in your choice to work on change and growth. It is not easy to be called out in the moment, so this one takes some careful planning to make sure it works for you.
- Starting a mindfulness practice is the way forward when you want to get a bird's-eye view of what your thoughts are doing. Having a set time of the day when you pay attention to your thoughts is a great idea. It's your formal practice to build that ability to step back from your thoughts and observe them without judgement.

A few pointers

As we are building awareness of our thoughts, we need to work hard to see that pattern of thought as just one possible interpretation of the world and allow ourselves to consider alternatives. Spotting these common thought biases and labelling them helps us to do that.

This is not something we do just once. It takes continued effort and practice. Sometimes you may not spot a thought as biased. At other times you will be able to spot them and come back at them with a more helpful alternative.

In trying to find alternatives, some people try to look for the correct answer. It is not so much the exact wording of the alternative perspective that is important. What matters more is the practice of stopping before you buy into a thought as fact and actively considering other views. As a general rule, it helps to look for a perspective that feels more balanced, fair and compassionate and that takes into account all the information available. Emotions tend to drive more extreme and biased views. But life is often more complex and full of grey areas. It's OK not to have a clear opinion on something while you take time to think about different sides of the story. So give yourself permission to sit on the fence for as long as you need to. Build up that ability to tolerate not knowing. When we do that, we are choosing to stop living life by the first thoughts that pop into our head. Our choices become more consciously thought-out.

Let's say I spill the milk all over the floor at breakfast and immediately start asking myself why I am such a failure in life and why nothing ever goes right for me. That's a nice mix of generalization and all-or-nothing thinking right there. If I can spot that bias and call it out, I can open up this window of opportunity in which I can reduce the intensity of that emotional response that might otherwise follow. It is never fun to spill milk, but our relationship with our thoughts can make the difference between a few minutes of frustration and something that ruins your mood for the entire day. Like everything in this book, it's very easy to say, but much harder to do. It takes practice and it doesn't make us invincible. But it helps, and it stops small moments turning into big ones.

Chapter summary

- Thought bias is inevitable but we are not helpless to its effects.
- We naturally look for evidence that confirms our beliefs. We then experience what we believe, even when there is evidence to suggest otherwise.
- Whatever has caused our low mood, it tends to come along with a focus on threat and loss (Gilbert, 1997).
- This bias towards the negative can then feed back to intensify the low mood if we continue to focus on and believe those thoughts to be facts.
- One strategy against the downward spiral this can cause is understanding that how we feel is not evidence that our thoughts are true.
- Another strategy is taking a stance of curiosity.
- Get some distance from those thoughts by becoming familiar with the common biases, noticing when they appear and labelling them as biases, not facts.

CHAPTER 3

Things that help

Getting some distance

In the 1994 movie *The Mask*, Jim Carrey plays a banker called Stanley Ipkiss. He finds a wooden mask that was created by Loki, the Norse god of mischief. When he puts it on, it wraps itself around the back of his head and consumes him, influencing his every move. He becomes the mask.

With the mask right in front of his face, he sees the world through that lens. There is no room for any other perspective. When he pulls it off his face and holds it in his hand at arm's length, the mask loses its power to change how Stanley feels and behaves. It is still there, but just that little bit of distance allows Stanley to see that it's just a mask, it's not who he is.

When we feel low in mood, thoughts can become all-consuming in this way. The brain senses from the body that things are not OK and starts offering up lots of reasons why that may be. Before you know it, a swarm of negative, self-critical thoughts are buzzing around your head. If we fuse with those thoughts and allow them to consume us, they can send the already low mood spiralling down further.

All those self-help books that told the world to just think positive didn't account for the fact that you can't control the thoughts that arrive in your mind. The part you can control is what you do once they appear.

One of the most important skills for learning to deal with thoughts and their impact on our mood is getting some distance from them. Sounds difficult when those thoughts are inside your own mind, but humans have a powerful tool that helps us to put thoughts at arm's length and give us the distance we need. It's called metacognition, which is a fancy name for thoughts about your thoughts.

We have this ability to think. But we also have the ability to think about what we are thinking. Metacognition is the process of stepping back from the thoughts and getting enough distance to allow us to see those thoughts for what they really are. When you do this, they lose some of their power over you and how you feel and behave. You get to choose how you respond to them rather than feeling controlled and driven by something.

Metacognition sounds complicated but it is simply the process of noticing which thoughts pop into your head and observing how they make you feel. You can have a go by pausing for a few minutes and noticing where your mind wanders to. Notice how you can choose to focus in on a thought, like

Stanley putting the mask over his face, or you can let it pass and wait for the next thought to arrive.

The power of any thought is in how much we buy into it. How much we believe it to be true and meaningful. When we observe our own thought processes in this way, we start to see thoughts for what they are, and what they are not. Thoughts are not facts. They are a mix of opinions, judgements, stories, memories, theories, interpretations, and predictions about the future. They are ideas offered up by your brain about ways we could make sense of the world. But the brain has limited information to go on. The brain's job is also to save you as much time and energy as possible. This means it takes short-cuts and makes guesses and predictions all the time.

Mindfulness is a great tool for practising the observation of your thoughts and strengthening that mental muscle that allows you to notice a thought and choose not to stick to it, but to let it pass, making a deliberate choice about where to focus your attention.

Mindfulness: Grabbing the spotlight

In the previous chapter I listed a few of the common thought biases that tend to be a feature of low mood. While some self-help books may tell you to 'just think positive', the problem with that is we cannot control the thoughts that arrive. If we try not to have certain thoughts then we are, by default, already thinking them. It's also not realistic. Lots of people face incredible hardship in their lives. We don't want to add to that burden by setting the impossible standard of only producing positive thoughts in terrible times. That is likely to add some self-criticism into the mix when they realize they cannot do that and start to think it is a personal failure.

So, while we cannot transcribe every thought that the mind produces, our power is in how we respond once it arrives.

When it comes to thoughts, attention is power. If you imagine your attention as a spotlight, many people leave that spotlight to move freely, wherever the wind blows. Your brain will occasionally take control if there are any signs of danger or threat. But we can also consciously choose to redirect that spotlight, paying attention to specific aspects of our experience on purpose.

This is not the same as blocking thoughts and trying to ignore them. It is being intentional about which thoughts you give the limelight to, which ones you zoom in on and turn up the volume.

Many people come to therapy knowing what they don't want. They know they have some thoughts and feelings that they would like to get rid of. But when we turn towards the future they do want, it can come as a bit of a shock, for the simple reason that they have never asked themselves that question before. Painful problems can be so overwhelming and demanding of our attention that we start to focus more on that and less on what we want instead.

Many of us are out of the habit of asking ourselves what we want. We have responsibilities: a boss to answer to, a mortgage to pay, children to feed. Over time, we come to realize that our wellbeing is not where we'd like it to be, but we have no idea what we actually want or need, because, well, we never think about it.

Now I am not about to tell you that you can manifest things in your life just by focusing on them. But we have to look where we are going if we want to stay on course.

Your attention is valuable and helps to create your experience of life. So learning to control where you direct it can have a powerful impact on your life and your mood. But we are busy, and life is full of daily responsibilities and duties that we have done a thousand times before. So our incredible brains like to make things easier for us by switching on to autopilot and doing most things automatically. That is why practices such as mindfulness meditation have become so popular. It enables us to get some formal practice. If you want to learn to drive, you take lessons. I see mindfulness practice as the driving lesson for managing the mind. It might feel boring, scary or frustrating at times, but it gives your brain the chance to lay down that neural pathway, so that when you need to use those skills later, they come without too much effort.

Practising mindfulness can be really daunting when you first get started, because you're not sure what you should be doing, if you are getting it right, or how you should feel while you're doing it. So in the toolkit at the end of this chapter you will find a few simple steps to guide you. It doesn't have to be complex. It doesn't have to be a profound experience. It's like lifting weights in the gym, but this time we are working a mental muscle, and as it grows, your ability to choose where you focus your attention increases, and along with it, so does your ability to manage your mood.

How to stop the rumination

Rumination is like a thoughts washing machine. It's the process of churning thoughts over and over for minutes, hours or days at a time.

We already know that the depressed brain is more likely to focus on the thought biases that can make you feel worse. If you combine those thought biases with the psychological equivalent of rumination, then you have a recipe for more intense and prolonged distress. In fact, we know from the research that rumination is a key factor in maintaining depression (Watkins & Roberts, 2020). The more you ruminate, the more you stay stuck. It works to intensify and prolong any sadness or depression that may be there.

Remember what we said before about neural pathways? The more you do something, the more established that neural activity becomes. This means the more you churn over painful thoughts or memories, the easier it becomes for you to bring those things to mind. You find yourself in a trap in which you continuously re-trigger painful emotions and distress and spiral down into a dark place.

So, what can we do to stop the rumination that feeds painful emotion?

When trying to change something in the moment, purely using a mental concept in our heads to re-focus on something new can be hugely difficult. I have seen many people use an active approach to good effect. When you notice that you are sliding down the slippery slope of rumination, try a firm hand pushed out in front and one word, 'Stop!', quickly followed by physical movement, such as standing up and moving away from the position you are in. Change activity for a moment, or even just walk around or step outside for a few minutes, whatever is possible at the time. Physically moving your body can help to shift your mind when it is otherwise very difficult.

Given that rumination invites us to swim around in thoughts of our worst traits and worst moments, and given the physiological implications of that on how we feel, one of the simplest ways to redirect things, when we are not sure of the way out, is a question: 'What would I do if I was at my best?' Now, if you are experiencing dark times and depression, you cannot expect yourself to be doing whatever you would be doing at your best. But you can create a mental picture of the direction you want to move in. So, if I am sitting ruminating on a painful experience in my life and have lost several hours to that churning, I can ask myself that question. The answer may be, 'I would stand up, take a shower and put some music on that lifts my mood. Or perhaps I would pick an activity I enjoy that absorbs my attention.'

For anyone who is prone to rumination, time alone opens up the gates of the arena for the thoughts and memories and subsequent emotional pain to come flooding in and start circling your mind. Human connection is quite possibly the most powerful tool we have to let those thoughts exit after a few rounds. Friends or a therapist will listen carefully to each one. But they can be great at holding up a mirror to our minds by reflecting back to us what they notice. They help to build our self-awareness and provide prompts or cues to call a stop to the rumination and shift to something new and more helpful to our wellbeing.

Mindfulness

Mindfulness is a state of mind that we can try to cultivate at any time. It means paying attention to the present moment, with awareness of thoughts, feelings and bodily sensations that arrive, without judgement or distraction. It does not rapidly eliminate low mood or change the problems you face. But it does hone your awareness of the details of your experience so that you are more able to choose carefully how you respond. But it can be difficult if you are not sure how to do it. Meditation is like a gym workout for the mind. It provides a space to practise the skills being used.

How to do it

If you are new to mindfulness, guided meditations are a good place to start. There are lots to choose from online, and I have included some on my YouTube channel. There are lots of different techniques, each based on their own traditions, but most of them share the purpose of aiming to create clarity of mind. So try out a few different styles and see which one suits you best.

Gratitude practice

Gratitude practice is another simple way to get used to turning your attention. Find a small notebook and, once a day, write down three things that you feel grateful for. They can be something big like your loved ones, or it can be a small detail of your day that you appreciate, such as the taste of your coffee as you sit down to work. Now, this sounds almost too simple to be effective, but every time you engage in gratitude, your brain is getting practice at turning its attention to things that create pleasant emotional states. The more practice you get – the easier it becomes to use that in other situations.

 Toolkit: Making gratitude a habit

- Write down three things you feel thankful for. They can be the larger, more profound aspects of your life or the tiniest of details from your day. What matters is not what you choose to include, but the practice of turning your attention on purpose.
- Spend a few minutes reflecting on those things and allowing yourself to feel the sensations and emotions that come along with a focus on gratitude.
- Doing this once is nice. Doing it every day is a life practice that builds the mental muscle to choose where you focus your attention and allows you to experience the benefits.

Chapter summary

- We cannot control the thoughts that pop into our minds, but we do have control of our spotlight of attention.
- Trying not to think about something tends to make us think about it more.
- Allowing all thoughts to be present, but choosing which ones we give our time and attention to, can have a powerful impact on our emotional experience.
- Turning our attention is a skill that can be practised with both mindfulness and gratitude practice.
- While there is a time for focusing on a problem, we also need to focus on the direction we want to move in, and how we want to feel or behave.
- Thoughts are not facts. They are suggestions offered up to us by the brain to help us make sense of the world.
- The power of any thought is in how much we believe it to be the only truth.
- Taking power out of those thoughts starts with stepping back, getting some distance (metacognition) and seeing them for what they are.

CHAPTER 4

How to turn
bad days into
better days

When we feel low in mood it can start to feel difficult to make decisions that, on better days, we make in an instant. Shall I call in sick or shall I push through and see how it goes? Shall I call my friend or shall I wait until I feel more up to it? Shall I try to eat something healthy or shall I eat something more comforting?

The problem with decision-making when you feel down is that low mood gives us the urge to do things that we know will keep us stuck. But the things that we know could help feel overwhelming. We start to focus on what is the best decision to make and berate ourselves for not doing it already. This is perfectionism rearing its head. Perfectionism paralyses the decision-making process because every decision has some inherent uncertainty. And every choice includes some negative side-effects to be tolerated.

When it comes to tackling low mood, we have to focus on making good decisions, not perfect decisions. A good decision is one that moves you in the direction you want to go. It doesn't have to catapult you there.

But something we must do is keep making decisions, however small. In any survival situation, making decisions and moving is essential. If you find yourself in deep water in the dark, with no way of telling which direction leads to safety, what you do know is that if you don't choose a direction and start moving, you won't be able to keep your head above the surface for long. Low mood wants you to do nothing. Therefore, doing anything positive, however small, is a healthy step in the direction you want to go.

Often what makes decisions more difficult in periods of low mood is our tendency to make decisions based on how we feel and how we want to feel right now. Basing our decisions instead on personal meaning and purpose can offer a shift from an emotional focus to decisions and actions based on our values. When dealing with low mood, focus on your personal values around health. What is important to you about your physical and mental health? How do you want to be living your daily life in a way that expresses that? How much are you living in line with those values at the moment? What is one thing you could do today that would steer you in the direction of looking after your health in the way you want to?

Consistency

When your mood is low and the small daily tasks feel like too much, don't go setting yourself extreme goals that feel out of reach. Pick one small change that you know you can action every day. Then make a promise to yourself that you will make it happen. It may seem silly at first, because those small changes won't reward you with instant, drastic results. But they are doing something much more important. They are laying down the pathways for a new habit that you can integrate into your everyday life and build upon over time as it becomes second nature. So keep it small. Keep it consistent. Slow change is sustainable change.

Don't kick yourself while you're down

We can't talk about tackling low mood without talking about self-criticism and self-attack. Low mood increases any self-criticism or self-attack that we already do. It's very easy to tell someone to just stop being hard on themselves. But when something has been a habit from a young age, just telling yourself to stop is not likely to cut it. We can't stop those thoughts arriving. But we can build our ability to notice them and respond in a way that gives them less power over how we feel and behave. We can use the same skills that we covered for catching thought biases and getting some distance from them. This helps us to recognize that those thoughts are emotion-loaded judgements, not facts.

Bring to mind someone you love unconditionally. Now imagine they were speaking about themselves the way that you speak to yourself. How would you respond to them? What would you want them to have the courage to see in themselves? How would you want them to speak to themselves instead?

This task is one way to help us access that deep sense of compassion that we often show to others but neglect to show for ourselves.

Self-compassion does not have to be airy-fairy self-indulgence. It is being the voice that you most need to hear, one that will give you the strength to pull yourself back up rather than drive you further into the ground. It is a voice of honesty, encouragement, support and kindness. It is the nurturing voice that dusts you off and looks you directly in the eye, telling you to go back out there and have another go. It is the parent, the coach, the personal cheerleader. There is a good reason that elite athletes have someone in their corner between rounds, sets and heats. They understand the potent impact of the words that fill your head. Whether you are in a boxing ring, a tennis court, a meeting at work or an exam hall, the same rule applies.

So talking to ourselves in the same way we would support and encourage someone we love is a powerful component in managing mood.

How do you want to feel instead?

When we are trying to tackle low mood, the tendency is to focus on everything we *don't* want to think and feel. There is value in doing that. But if we want to move away from what we *don't* want, it helps to know where we *do* want to go instead.

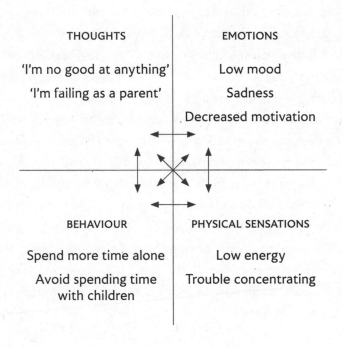

Figure 5: Example formulation for low mood.

Toolkit : Figuring out what you can do to change how you feel

Start by filling in the hot cross bun formulation for the low mood scenario found at the back of the book (see page 347). I've put an example on the previous page (Figure 5).

Once you have broken down the thoughts and behaviours that are contributing to your low mood, fill in the blank formulation for your better days (see page 348). This time, start with the emotions box and fill it with the emotions you would like to feel more of in your everyday life instead of the low mood. I have put an example on the next page (Figure 6).

Knowing that your physical state, the focus of your thoughts and your behaviour all contribute to that feeling, use the following prompts to help you fill in the rest of that formulation:

- When you have felt that way in the past, what has been the focus of your attention?
- What might your thoughts/self-talk need to sound like in order for you to feel that way?
- When you have felt that way before, how did you behave? What did you do more of or less of?
- If you were to feel that way, how would you need to be treating your body?
- When you are feeling at your best, what do your thoughts sound like?
- What do you tend to be focused on? What does that inner voice sound like then?

49

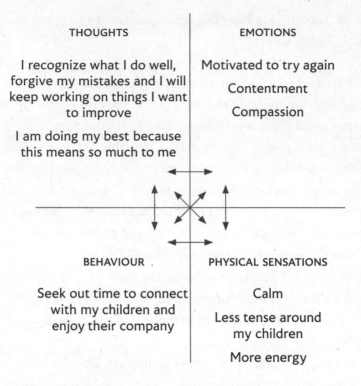

THOUGHTS | EMOTIONS

I recognize what I do well, forgive my mistakes and I will keep working on things I want to improve

I am doing my best because this means so much to me

Motivated to try again

Contentment

Compassion

BEHAVIOUR | PHYSICAL SENSATIONS

Seek out time to connect with my children and enjoy their company

Calm

Less tense around my children

More energy

Figure 6: Example formulation for better days. How do you want to feel, behave, think?

See if it reveals anything that has worked for you in the past, or offers insights into things that you could pay attention to or change about your daily life. Use it to spend time examining what works.

 Try this (solution-focused miracle question): Take a moment to imagine that when you close this book a miracle happens and the problems you have been struggling with all disappear.

- What would be the first signs that the problem has gone?
- What would you do differently?
- What would you say yes to?
- What would you say no to?
- What would you focus your energy and attention on?
- What would you do more or less of?
- How would you interact with people differently?
- How would you structure your life differently?
- How would you speak to yourself differently?
- What would you be free to let go of?

Take some time to explore your answers to these questions, down to the finest detail of those small changes that you would make to your everyday life. This is a great exercise for creating a vision of where you are heading. It also helps you to explore the idea of how life might be improved by starting to make some of those changes now, even with the problems still present. What we do and how we do it feeds back to our body and brain about how to feel, so shifting direction towards what matters most, and the person we want to be alongside our problems, can bring about big shifts in mood. This technique shifts our focus from the problem to the solution, and we can start to fix our gaze on the horizon that we are heading towards.

Chapter summary

- Focus on making good decisions, not perfect ones. 'Good enough' steers you towards real change. Perfectionism causes decision-making paralysis, whereas improving your mood demands that you make decisions and take action.
- Keep changes small and sustainable.
- When someone is down, we show them kindness because we know it is what they need. So, if you are committed to managing your mood and overall mental health, commit to practising self-compassion.
- Once you understand the problem, use it to help you work out where you want to go and focus on the horizon ahead of you.

CHAPTER 5

How to get the basics right

Imagine taking the best football team in the world and putting them on the pitch without any defence players on their side. Suddenly, opponents that never previously posed any threat would stand a much better chance at winning. While the defence players may not be as exciting as the strikers, we all underestimate their power to change the game.

We all have the tendency to neglect the basics. Your mother tells you to get to bed early and eat your greens and you can't help but roll your eyes. Thanks, Mum. But when it comes to defence players, you don't know what you've got till it's gone. The basics are the first thing we let slide when we are not feeling so good. We withdraw from friends, we drink too much coffee and then can't sleep, we stop exercising. But what difference does it really make? Well, the science suggests that it's a bit like taking each of your defence players off the pitch and leaving an open goal.

The basics are not glamorous. They don't give us that hit of having bought something that promises to fix everything. But they are cash in the health bank. When life starts throwing things at you, those defences are going to keep you standing, and help to pull you back up if you fall.

It's worth pointing out that you don't have to approach them all perfectly. There is no perfect diet that everyone agrees on, or optimal amount and type of social interaction. These are not goals to be achieved to perfection. They are foundations. Those defence players need as much nurturing as the strikers, because they are crucial for staying in the game. But where one slips, another can help out. When those defences get compromised, it's not a sign of fault or failure. It's generally a sign that life is happening. For example, a new parent is likely to have no control over the sleep deprivation they face. But they can take extra care to eat well and keep up social contact with friends and family, which will help to keep them well throughout that period.

Understanding what those main defences are means that we can keep them on the radar. We can check in with them on a regular basis and look for small ways in which we might be able to improve them and strengthen them.

If you are tempted to skip over this because you think you've heard it all before, then it's even more important to read it. We underestimate the power of these defences so much that they are often the first things we let go when we are under stress or not feeling so good. But not only is the science crystal clear on these; in recent years, it has also shown their potency to be even more far-reaching than we once thought.

Exercise

Whether your low mood is mild and intermittent or you live with a major depressive disorder, exercise has potent antidepressant effects (Schuch et al., 2016). For those who use antidepressant medication, adding in exercise leads to better results (Mura et al., 2014).

Exercise leads to higher circulating levels of dopamine as well as more available dopamine receptors in the brain (Olsen, 2011). This means it increases your capacity for pleasure in everyday life (McGonigal, 2019). So finding exercise you enjoy does not only offer you joy while you are exercising, but increases your sensitivity to find joy in all the other aspects of your life.

Unfortunately for us, the concept of exercise has been branded as a painful process that you must go through for the purpose of altering your appearance. The conversation is mostly centred around enduring pain for some aesthetic gain. No wonder that so many people see exercise as not for them.

The focus on how moving your body makes you *feel* has been left out of the conversation for too long. Yet, throughout the pandemic many people have rediscovered the joys of exercising in green spaces. Being forced to spend so much time indoors made the effects of that daily walk all the more salient. The psychological impact of exercising, not on a treadmill indoors, but outside in nature, is starting to be demonstrated by the science. In a study of adults undergoing Cognitive

Behavioural Therapy (CBT) for depression, the group whose therapy was provided in a forest setting showed a remission rate that was 61 per cent higher than those who underwent the same programme in a hospital setting (Kim et al., 2009).

For those who can't stand the idea of vigorous exercise, the slower movements of yoga can still have significant effects on mood as well as increase the ability to calm the mind and body more quickly (Josefsson et al., 2013).

Deciding to add in exercise to your life does not mean you have to take up ultra marathons or weightlifting at an expensive gym. In fact, it is much easier to gain momentum if you start as small as you can. You might not even leave the house to begin with. You might put on your favourite music and dance around until you feel a little out of breath. If you can make it small and choose something that has the potential to bring you joy, it is much more likely to be sustainable. A one-off workout won't change everything, but a small increase in physical activity that you can keep up has the power to become a catalyst for significant life change.

Exercise does much more than just give you a little boost in mood. It positively impacts on your mind and body in countless ways. But don't take my word for it. Find a way to increase your activity in a way that feels enjoyable or meaningful to you and see how you start to feel.

Sleep

Take anyone on this earth and start depriving them of sleep and that person will become vulnerable to both physical and mental illness. But that relationship between sleep and mental health works both ways. When your mental health dips because of stress, low mood or anxiety, your sleep is likely to be disturbed at some point too. Whichever came first, you can almost guarantee that when your sleep takes a dive so will your mood and your belief in your ability to bounce back. When you haven't had enough sleep, everything feels ten times harder. Sleep has a profound effect on every aspect of your wellbeing, so if you think your sleep is not as good as it could be it is well worth your time and effort to try improving it.

For those who experience long-term insomnia, working with a specialist is highly recommended. But if you want to work on improving the amount or quality of sleep you are getting, I have included a list of tips below to get you started. Again, we are not aiming for perfection and you don't have to do everything on the list to get good enough sleep. There are times when life happens and it pulls you away from a healthy sleep pattern. If you work shifts, travel long-haul, have young children or a habit of staying up late playing computer games, you can learn to check in with yourself and put in place whatever is necessary to steer you back in the right direction.

- Keep vigorous exercise to earlier in the day and try using the evenings for relaxation.
- A warm bath before bed can help your body reach a temperature that is optimal for sleep.
- Try to get as much natural light as possible within the first 30 minutes after waking. Our circadian rhythm that regulates our sleep pattern is governed by exposure to light. Indoor light can help, but natural light outdoors is best, even on a cloudy day. Step outside for 10 minutes first thing in the morning. Also make time to get outside as much as possible throughout the day.
- In the evening, when the sun goes down, keep lights low. When it comes to screens, the research suggests it is less about the colour of the screen light and more about the brightness. So turn down the brightness on screens in the evening as much as possible and switch them off altogether as early as possible.
- Make time in the day to tackle things you are worried about. Make your decisions, make your plans, take something off your to-do list. Getting good sleep is also about what you do during the day. We are much better at problem-solving during the daylight hours, but if we push those problems away and ignore them, they tend to pop up at night when we are trying to sleep. So clear the desk and clear your mind as much as you can.
- For those nights when laying your head on the pillow is the cue for your brain to switch on and start worrying, try keeping a worry list. Put pen and paper beside the

bed. When a worry pops into your mind, write it down. Just a few words or bullet points. Do the same for any other worries that come up. This becomes your to-do list for the next day. You make a promise to yourself that you will dedicate time tomorrow to working through these problems. That way you are free to let them go for now and re-focus your attention on rest.

- You can't force sleep. Making sleep happen is not something you can choose to do. Sleep happens when we create an environment in which the body and mind can feel safe and calm. So don't concentrate on sleep, concentrate on relaxation, rest and calm. Your brain will do the rest.
- Avoid caffeine in the late afternoon and evening. The energy drinks that are being promoted to young people often contain high levels of caffeine that disturb sleep and cause symptoms of anxiety.
- As a general rule it's a good idea not to consume too much of anything in the lead up to bedtime. This especially includes big meals with high levels of refined sugar. Anything that gives you a spike in stress levels is not going to help you get to sleep and stay asleep.

Nutrition

Mental health and physical health are weaves in the same basket. If one moves, the other moves. In recent years the science has made strides in demonstrating this. How you feed your brain influences how you feel.

The research is even showing that improvements in nutrition can have large benefits for depressive symptoms (Jacka et al., 2017) and making positive changes to how we eat might help prevent depression as we age (Sanchez-Villegas et al., 2013).

When we understand that our mood is influenced by several factors, it makes good sense to tackle that from all sides. With just a moment of reflection, most of us could easily come up with a few ways in which we could nourish our body better. The research being done across the world suggests there is not one, strict diet that protects your mental health. The traditional Mediterranean diet shows the largest and strongest evidence base for mental health benefits, but many others show a reduced risk of depression, including the traditional Norwegian, Japanese and Anglo-Saxon diets too (Jacka, 2019). The things they all tend to have in common are the inclusion of whole, unprocessed food, healthy fats and wholegrains.

There is a lot of misinformation around food, so in the resources section of this book I have included suggestions for further reading from credible sources, if you would like to read further. But the overarching idea is that making good nutrition a priority (and educating yourself about what that looks like if you need to) is a great idea for tackling low mood and improving your mental health.

But as I mentioned before, making huge life transformations overnight is less helpful if you cannot sustain them. Instead, it is helpful to ask ourselves on a regular basis, 'What is one small change I could put in place today that would improve my nutritional intake?' Then repeat this every day.

Routine

Another key defence player for mental health and resilience seems to be routine. This may have been the most underestimated influence on our wellbeing until the pandemic turned routine upside down and inside out for so many people.

Repetition and predictability help us feel safe, but we also have a need for variety and a sense of adventure. So we like to have routine, and we like to break those routines occasionally – preferably with something pleasurable, meaningful or exciting.

When we are not feeling so great, routine can suffer. You might stay up late watching television because you are blocking out stressful thoughts about work the next day. Then it's harder to get up in the morning, so you stop your morning workout.

Or maybe you are out of work for a period of time and you start taking a nap in the afternoon, but then struggle to sleep at night. Being off work changes how much social interaction you get. You don't leave the house for days and struggle with finding a reason to shower, or even get up in the morning. Then your appetite goes but you have no energy, so you mostly spend the day drinking coffee . . . and so the cascading effect of a routine change unfolds.

Each of these seemingly small changes matters because they add up to and create your experience as a whole. If you have a tall glass of water and you pour a small drop of cordial into it, you might barely be able to notice a difference. If you add another couple of small drops, the general colour of the water begins to change. Enough of those small drops over time will dramatically change the overall colour and taste of the water. So each small drop counts, even though they may not be enough by themselves to change how you feel entirely.

That said, there is no perfect routine. Establishing a balance of predictability and adventure that works for you within your unique circumstances is key. Noticing when that goes off track and pulling it back is a big step in the right direction.

Human connection

While looking after your own body and mind is essential, nurturing good quality relationships is one of the most powerful tools we have in maintaining good mental health throughout the lifespan (Waldinger & Schulz, 2010).

When our relationships are not going well, it can have a catastrophic impact on our mood and emotional state. It can also work the other way. Deterioration in our mood can be damaging to our relationships. It can leave us feeling disconnected from the people around us and trigger a deep sense of loneliness.

When you're feeling low, the thought of facing anyone can be exhausting and overwhelming. This is the trap of depression. It tells us to withdraw, hide away, don't see anyone until we feel better. So we wait to feel better. But in doing so we hold ourselves back. Taking some time to be alone can be re-energizing and recharging, but it can also very easily become a downward spiral of rumination and self-loathing that feeds depression and keeps it going.

Being with others (even when we don't feel like it), to observe them, interact with them and build connections with them, can help to lift our mood and pull us out of our own head and back into the real world. We know from the research that good-quality social support is associated with better outcomes when it comes to mood (Nakahara et al., 2009).

Many of those people who never tell a soul when they're struggling firmly believe that presenting themselves as anything less than their best would make them a burden to the people around them. But the science suggests otherwise. Social support has positive effects for both the one receiving it and the one providing it (Inagaki et al., 2012). So, when we are struggling and we want to pull ourselves up from low mood, one of the most powerful things we can do is swim against that strong tide that is pushing us towards isolation and loneliness. We must not wait until we feel like it, because feeling like it doesn't come first, the action must come first. The feeling follows on after. The more time you spend making real connections with other people, the more you will start to improve your mental health.

Spending time with other people doesn't mean we have to talk about how we feel. In fact, we don't have to talk at all. Just be around people, watch them, smile at them. Share whatever conversation you can manage. Low mood and depression can make us feel uncomfortable and anxious around people. We get concerned with how we might be coming across. We tend to spend so much of our time criticizing ourselves that we start to assume that others are judging us too. Can you remember which thought bias this was?

Despite all the thoughts and feelings that hold us back from each other, human connection is our inbuilt mechanism for resilience. When we struggle, connection helps. Good-quality, safe connection. If that can't be found in family or friends, then professionals can provide that until you are able to find and create new, meaningful relationships in your own life.

Chapter summary

- Our mental health defence players provide the foundations of good health. When we nurture them daily they pay us back with interest.
- If you do one thing today, make it exercise. Choose something you enjoy and you increase your chances of keeping it going.
- The relationship between sleep and mental health works both ways. Prioritizing sleep will help your mental health, and making changes to your day will affect your sleep.
- How you feed your brain influences how you feel. Traditional Mediterranean, Japanese and Norwegian diets show benefits for mental health.
- Human connection is a powerful tool for stress resilience. Your relationships change your biology and psychology.

On
Motivation

CHAPTER 6

Understanding motivation

As we build up the psychological toolkit with skills that help us manage through life, it's easy to imagine that motivation is one of those tools. But motivation is not a skill. Neither is it a fixed personality trait that we are either born with or without.

So many of us know exactly what we need to do, we just really don't feel like doing it right now. And when later comes, we don't feel like it then either. Sometimes we can get all fired up about a goal and things start moving in the right direction. But a few days later that feeling fizzles out again and we're back to square one.

Motivation that goes up and down is not a fault in the system. It's part of being human. It is a sensation that comes and goes just like our emotions, so we can't always rely on it to be there. But what does that mean for our dreams and goals?

Your brain is constantly paying attention to what is happening in your body. It knows what is happening to your heart rate, your breathing, your muscles, and it reacts to the information it receives, making judgements about how much energy it should expend on the task in front of you. This means we have more influence over those feelings than we think. When we start to change what we are doing with the body, it has an influence on the activity in the brain, which in turn influences the feelings that are produced in the body. This is something we can use to our advantage.

When we're dealing with that 'can't be bothered' feeling there are two main prongs of attack:

- Learning how to cultivate that feeling of motivation and energy to increase the chances of it appearing more often.
- Learning how to act in line with your best interests even when motivation is absent. Developing the capacity to do what you need to do, even when a part of you doesn't feel like it.

Procrastination or anhedonia

I want to make a distinction here between procrastination and anhedonia. Procrastination is something everyone does. It's when we put something off because the job we need to do triggers a stress reaction, or some other feeling that is aversive. I have made hundreds of educational videos for social media, but give me a video to make that feels uncomfortable or difficult to get right and I will skirt around the job all day, convincing myself that all those other things I am getting done is me being productive. In fact, it's usually me procrastinating because the idea of making that particular video feels difficult or uncomfortable that day.

Anhedonia is something different. This is when we stop taking pleasure in the things that we used to enjoy. Anhedonia is associated with a number of mental health problems, including depression. When we feel that way, we start to question whether anything is worth the effort. Things that once brought joy start to feel meaningless. So we stop doing the things that have the potential to lift our mood because we have no desire for them any more.

When you start to avoid something that is important or potentially meaningful to you, the natural response is to wait until you feel like it once again. You wait until you feel energized or motivated or ready. The problem with this is that the feeling does not arrive spontaneously, we need to create it through action. Doing nothing feeds the lethargy and that 'can't be bothered' feeling and makes it worse. Motivation is a wonderful by-product of action. It's that great feeling you get when you are on your way out of the gym, not on your way in. It's that feeling of energy and momentum you get once you have started something and your brain and body start to rise to the challenge for you. Sometimes the feeling is fleeting. At other times it lasts for much longer. Much of that will depend on all the other factors that are either working to foster it or squash it.

So when we start doing something, even when your flat mood says, 'I don't feel like it,' we can trigger a biological and emotional shift. This doesn't mean that putting some music on or doing a single workout is going to solve all your problems or change your life. But it sets in motion a series of events that shifts your direction. If you start doing that thing that you wish you felt like doing, you have more chance of stimulating your brain in a way that brings about enjoyment or a sense of motivation.

For someone who is struggling with depression, and suffering from anhedonia as a part of that, the pleasure in activities and the motivation to engage in them takes time to return and can be very up and down for a long while. There is a period in which we have to grind, doing things that matter to us, even when we don't feel like it, in order to re-engage with the pleasure we used to feel.

Chapter summary

- Motivation is not something you are born with.
- The feeling that you are energized and want to do something cannot be relied upon to always be there.
- Mastering motivation is building the capacity to do what matters most to you, even when a part of you does not feel like it.
- Procrastination is often avoidance of stress or discomfort.
- Anhedonia is when we no longer get a sense of pleasure from the activities that we used to enjoy. This is often associated with low mood and depression.
- If something matters to you and could benefit your health, don't wait until you feel like it – do it anyway.

CHAPTER 7

How to nurture that motivation feeling

Motivation is more than just a reason for doing something. When we use that word in conversation, we often mean a feeling of enthusiasm or drive that fluctuates just like any other. Some things nurture that feeling, others flatten it. What things do you do that are often followed by a feeling of motivation and energy?

Science tells us about the things that work for most people. But the detail you can establish from looking at your own life with curiosity adds significant value. You cannot change something you are not aware of. So spending time observing and documenting the thing you are trying to tackle is hugely important and gives you the best chance of generating that feeling of motivation more of the time.

Here are some things that invite that feeling to be present:

Move your body

Motivation does not come from a specific location in your brain. It is not a fixed part of your personality. It's also not an essential tool that we use to make us move. It is most often a consequence of that movement.

But what if you have no motivation to exercise? Perhaps the key to making exercise a sustainable part of everyday life is to find a form of movement that you can begin even when your motivation is low. Research shows that doing even small amounts of exercise is better than nothing, and anything more than your usual amount of movement will help boost your willpower (Barton & Pretty, 2010). Find something that feels easy. Something that brings you joy. Something that feels like precious time out rather than another boring job that has to get done. Add in friends, good music, and anything that helps you to look forward to it each day, rather than dread it.

Adding in some form of exercise, however moderate, will pay you back in feelings of motivation. You may have to use the strategy of acting opposite to the urges, because you won't feel like working out. But the impact this simple action will have on that 'can't be bothered' feeling for the rest of the day is unmatched. Make this one thing happen and you are setting yourself up to win.

Staying connected with the goal

In therapy we often set goals with people and help them to work out how they are going to achieve them. But the real work happens when things slip off track. This is the point at which those without support might be vulnerable to giving up. But we need to get to work using the setback to strengthen the future. If we can better understand what caused the failure and that getting back on track is just part of the process, then we are in a good position to predict when it could happen again and steer around those challenges in the future.

I think one of the reasons that some of my clients say they feel so much more motivated after an appointment is because they have spent time reconnecting with their goals. If that thing we are working on is not fresh in our minds we can quickly lose momentum.

Whether you are working on improving mood or any other aspect of your wellbeing, it is paramount that you stay connected with your goals because they will demand constant nurturing. Return to them on a daily basis. You can do this through journalling. It doesn't have to be a hugely time-consuming task. It can be one minute at the beginning of your day, listing the one or two things you will be doing that day towards your goal. Then, at the end of the day, writing a few lines to reflect on your experience. This kind of task is easy to maintain because it doesn't require too much time – maybe a couple of minutes at most. But it ensures that you are accountable to yourself every day and keeps you focused on your goals.

Keep it small

Any big task will invite that 'can't be bothered' feeling, so keep it small and keep it focused. People transform their lives with therapy but that doesn't happen overnight. They don't return to their second appointment problem-free and with an entirely new mindset. Individuals take home just one task at a time and focus on that. We can only focus on one thing at a time and we only have limited ability to do things that we don't feel like doing.

But of course most of us don't stick to this. We can see that life needs an overhaul and we try to do one big transformation all at once. We expect too much of ourselves and then fall into despair when we burn out or give up. When that happens, we are less likely to try again.

When motivation for a long-term goal dips, it helps to have small rewards along the way. Not so much external rewards, but internal ones. That emotional pat on the back you provide for yourself when you congratulate yourself for your efforts and acknowledge that it has been worth it because you are heading in the right direction. Doing this helps you to lean back into effort, knowing that you are on the way towards the changes you want to see.

When we acknowledge the progress and small victories along the way, we start to recognize that our efforts can influence our world. Feeling that we have agency in this way helps us to feel energized to keep trying. This is a great reason to start small and develop new habits, ensuring that each one gets embedded. Once you sustain the habit of prioritizing your healthy behaviours, they will sustain you.

Resisting temptation

Sometimes we are trying to build motivation to help us take action. But change can also demand the willpower to resist temptation and the urges to do things that take us in the opposite direction to our goals.

I must have been around the age of three or four when I visited my grandparents' house and walked into the garden to find my grandfather using a garden strimmer. Something wasn't working right, so he flipped the tool upside down and started pulling bits of grass out from between the blades. He turned to me and said, 'Whatever you do, do not press that red button!'

I sat on the grass next to the red button on the side of the strimmer and fixed my gaze on it. *Don't press it. Don't press it.* I wondered if it was one of those buttons that gave a satisfying *click* when you pressed it. *Don't press it.* It looked really smooth along the top. *Don't press it.* Like a magnet, my hand reached out and pressed that red button. The strimmer instantly responded with a loud noise as the blades swung into action. As luck would have it, no fingers were lost that day, but I did learn a new curse word.

Focusing on the thing I was not supposed to do turned out not to be a helpful strategy. So what *does* help when positive change demands that we resist temptation? One of the biggest factors is managing stress. The physiology of self-control is optimal when stress is low and heart-rate variability is high. Heart-rate variability is a measure of the variation in time between each heartbeat. So it tells us how much your heart rate changes throughout the day. You might notice that when you get out of bed in the morning your heart speeds up, or when you run for the bus. Then it gradually comes back down again. This means your body gears you up for action when needed and calms the body back down to rest and recover. But when we are under lots of stress, heart rate may remain high throughout the day (reduced variability).

When it comes to resisting temptation and maximizing our willpower, we need that capacity to calm the body and mind. Anything that increases stress is going to have a negative impact on our ability to make wise choices for our future. Stress increases the likelihood that we will instead act based on how we feel right now and sabotage our goals. So if you are sleep-deprived, depressed, anxious or not eating well, your heart-rate variability goes down, along with your chances of sticking to your goals. Whether you are trying to give up smoking or unhealthy foods, or just trying to regulate your emotions in healthier ways, when it comes to relieving stress and increasing your capacity for willpower, exercise is a top choice. It has both an immediate effect and a long-term impact (Oaten & Cheng, 2006; Rensburg et al., 2009).

So whatever change you are working on, increasing your level of activity, even in a small way, is a great way of strengthening your willpower to keep it up (McGonigal, 2012).

Another big player in managing stress and the ability to

make wise decisions is sleep. You only need one bad night's sleep to struggle with increased stress, trouble concentrating and low mood the next day. Self-control takes energy and if you haven't had enough sleep, your brain has less access to that energy and becomes more vulnerable to high stress reactions, squashing your capacity to control your actions.

Change your relationship with failure

Something that can zap motivation is the prospect of failure. But that depends on the relationship we have with failure. If slipping up and going off track means we engage in vicious self-attack and relentless self-criticism, we are likely to feel ashamed and defeated. If we associate failure with unworthiness, then starting anything new is going to feel overwhelming and procrastination will be front and centre. We protect ourselves from the psychological threat of shame by sabotaging the process before it gets started.

Shame is not as helpful for motivation as you might think. When we get caught in self-criticism and shame, we feel inadequate, defective and inferior. When we feel that way we want to hide, get smaller, disappear. It produces urges to escape and avoid, rather than to dust ourselves off and try again. In fact, it is so painful that it induces strong urges to block that feeling, which is risky for anyone living with addiction. So if we want to persist at something and feel motivated to keep trying, we have to think carefully about how we respond to failure along the way.

If there is ever resistance in therapy, it is when you explore the idea of being self-compassionate. I hear things like, 'I will lose my drive, become lazy.' 'I would never achieve anything.' 'I can't just let myself off the hook like that.' Most people are shocked and surprised to find out that self-criticism is more likely to lead to an increase in depression rather than motivation (Gilbert et al., 2010). Self-compassion, on the other hand, treating yourself with kindness, respect, honesty and encouragement after a failure, is associated with increased motivation and better outcomes (Wohl et al., 2010).

 Try this: If we are not aware of our self-criticism and the impact it is having on our fear of failure and motivation, then it is much harder to change it. Use these prompts to reflect on the way you talk to yourself after a setback.

- When you experience failure, what does your self-criticism sound like?
- What emotions are attached to it?
- Do you suppose that the failure reveals something about your inadequacy or incompetence as a person?
- Do you notice any shame or hopelessness associated with that?
- What coping strategies tend to follow that self-criticism?

- How does that impact on your original goals?
- Think of a time when you failed at something and someone responded to you with kindness and encouragement. How did that feel? How did that help you to try again and succeed?

 Toolkit: How to respond to failure with compassion and accountability so you can get back on track

Bring to mind a recent memory of a failure or setback. Then work through the following exercise.

1. Notice what emotions are brought up by that memory and where you feel them in your body.
2. How did the self-criticism sound? What words and phrases came up and how did they influence how you felt?
3. How did you then respond to the feelings?
4. Bring to mind someone that you love or respect. If they experienced the same failure, how might you have responded to them differently? Why would you have shown them that respect?
5. How would you want them to perceive the setback in order to get them back on track?

Chapter summary

- While we can't control the feeling of motivation, these are things we can do to increase the chances of feeling it more often.
- Physical movement cultivates feelings of motivation. Small amounts are better than nothing and can help build momentum.
- Staying connected with your goal helps to keep triggering moments of increased motivation.
- Small and consistent beats one-off grand gestures.
- Learning to rest and replenish between stressful situations helps to maximize willpower.
- Shame is not as helpful for motivation as you might think. Changing your relationship with failure will help your motivation.

CHAPTER 8

How do you make yourself do something when you don't feel like it?

No matter how much we reduce our stress and cultivate motivation, it can be fleeting. It comes and goes. So we cannot rely on it to always be there. And there will always be things that we will never feel like doing. Tax returns, insurance renewals, taking the bins out. How do we make those happen, even when a part of us would prefer not to?

Feelings are often accompanied by urges. Those urges are suggestions, nudges, persuasions telling us to try this or that to relieve the discomfort that we feel or to seek the reward that we anticipate. While those urges can be powerful, we don't have to do what they say.

Opposite action

When we were children, my sisters and I would share a packet of Polo mints and compete for who could go the longest without crunching the mint – a challenge that is much more difficult than it sounds. The urge to crunch the mints felt almost undeniable at the time. It demanded intense concentration and focus. As soon as you got distracted and let your guard down, your brain would take over on autopilot for you and the mint was history.

If you try that game, what you notice is that your awareness focuses on your experience. You get to observe the sensation of an urge. And you get to create a gap between urge and action. Simply by paying attention, you get to choose whether you go with the urge or against it. When the task is as simple as holding back from crunching a mint, all you need is a bit of competitive sibling rivalry to keep you on task. When we are trying to tackle much stronger urges toward ingrained patterns of behaviour that come along with intense emotional states, the challenge is much harder.

The skill of acting opposite to an urge, to instead choose a behaviour that is more in line with where you want to go, is a key skill that people learn in therapy (Linehan, 1993). The opposite action skill is the deliberate attempt to take an action that is the opposite of what the emotion is telling you to do.

This is especially helpful when your coping strategies tend to cause you harm.

Mindfulness is a key component of this skill. Paying attention to our experience, and the thoughts, emotions and urges that come with it, allows us to pause for just long enough to make an informed, sometimes pre-planned decision about what to do next. This means we can act based on our values rather than our emotions.

The pain barrier

The best strategy for motivation is to take motivation out of the equation. There are things we do every day, whether we feel like it or not. For example, in the morning, you don't ask yourself whether you have the motivation to clean your teeth, it's just something you do. It is a habit so well-practised that you don't need to think about it any more. You just do it. The reason for that is because it has been a non-negotiable part of your daily routine for most of your life.

Imagine your brain is like a jungle. For every action that you take, the brain has to make connections or pathways between different areas. When you repeat an action on a regular basis over a long period of time (like cleaning your teeth) those pathways become well-trodden and established. Those smooth, wide paths become easier to access so that your brain can do much of that action without you having to consciously think too much about it.

But when you start something new you have to carve out a fresh path, sometimes from scratch. That takes huge amounts of conscious effort. And if you don't use that path often enough, it will always be effortful. Any time that you are under stress your brain will automatically choose the easiest route, which is the path well-trodden. But if you can repeat that new behaviour as often as you can, for enough times, then a new habit is established and becomes easier to use when you most need it.

Here are a few tips on how to establish a new habit:

- Make the new behaviour as easy as possible to do, especially in moments when you might not feel like taking action.
- Set up your environment in a way that supports your new change of behaviour. In the early stages of the change, you cannot rely on habit.
- Make clear plans and set reminders if necessary.
- Add in a mix of short-term and longer-term rewards. Internal rewards work better than external ones. So we don't need trophies as much as we need that internal celebration and acknowledgement that we are on the right track.
- Get clarity on why you are making this change and why it matters so much to you. You can use the values exercises in this book to help with that (see page 286). Establish this change as a part of your identity. This is how you do things now.

How to persist for the long-haul

Over the years, psychology research has challenged the idea that success is all down to innate talents and has shown that grit (Duckworth et al., 2007) and, in particular, perseverance play a vital role in our ability to succeed (Crede et al., 2017). But how do we achieve the kind of stamina that we need to be able to persevere even in the face of setbacks?

Something that many people learn the hard way is that it doesn't mean just continuing to drive forward until you burn out. When we are working on long-term goals and making changes that we want to maintain, we have to learn to counter-balance the stress of effort with the replenishment of rest. We don't need to always be working or always feel energized and refreshed. We need to be able to listen to the body and step back from effort so that we are ready to drive forward once again.

Just as elite athletes may nap in between training sessions and professional singers rest their voices by going for days without speaking, we need to recognize that regular rest and replenishment is crucial if we want to persevere at anything for the long-haul.

But not all breaks are equal. Most days are broken up by those quiet, sometimes boring moments in between periods of intense work and effort. But if we use those in-between moments to clear emails, scroll through social media or get a few things done, the body and brain will not be returning to a rest state to recharge. So the next time you reach for your phone to fill the 15 minutes between meetings, why not step outside for some fresh air or find a space to close your eyes for a moment instead?

We also need to make use of small rewards as we work on big goals. When we break down big challenges into smaller tasks and reward ourselves for reaching those milestones, we get the benefits of small dopamine hits along the way. Dopamine not only gives us a little 'buzz' that feels rewarding, it also drives us to look ahead to the next milestone and motivates us to keeping driving forward. It enables us to imagine how we might feel once we accomplish the challenge faced, and triggers desire and enthusiasm (Lieberman & Long, 2019). So, building in those small rewards along the way helps to refuel your desire for the end goal and your capacity to persevere.

Let's say you are trying to run further than you ever have. When you start to feel tired, you tell yourself you are just going to get to the end of this road. When you do, you mentally give yourself a pat on the back for achieving it. You see it as a sign that you are on the right track. This internal reward to yourself gives you a dopamine release. This suppresses the noradrenaline that causes you to give up. As a result, you get an extra boost to keep going for a bit longer. This is not the same as positive self-talk. You focus on a small, specific goal and the accomplishment of it means you are heading in the direction of your ultimate goal (Huberman, 2021).

So, when the task ahead feels like a mountain to climb, you don't look up at the peak. You narrow your focus and set yourself the challenge to make it to that next ridge. When you get there, you allow yourself to absorb that feeling of being on your way. Then you go again.

Gratitude

A gratitude practice can be a powerful tool for longer-term goals that demand persistent effort. Turning your attention towards gratitude self-generates those internal rewards that replenish and restore your capacity to keep returning to the effort that is required of you. A simple shift of language can help us turn towards gratitude. For example, try switching 'I have to . . .' with 'I get to . . .'

As previously mentioned, we can also practise gratitude in a more formal way by sitting down with pen and paper and writing down the things that we feel grateful for each day. When we do that, we are turning our attention on purpose in a way that shifts our emotional state. But it is not only that immediate impact on our emotions that we gain. When we practise gratitude on a regular basis we are repeating an action. As described previously, the more we repeat an action the easier it becomes for the brain to do it with less effort in the future. Almost like a mental muscle, putting the reps in each day makes it much easier to think in a helpful way when we might need to in the future.

Pre-planning

In therapy we often create crisis plans with people. Sometimes those are about staying safe in life or death situations. At other times they are to prevent relapse of addiction or getting off track with a goal at a time when they might be vulnerable to giving up. You can make use of this to improve your chances of sticking to any plan. Look ahead at the change you want to make. Write down all the potential hurdles that could cause you to get off track. For each hurdle, make a plan of action about how you will prevent those hurdles from causing you to get off track or give up on your goals. Engineer the situation in advance, to make acting in line with your values and goals as easy as possible, and sabotaging those goals based on emotional urges as difficult as possible. For example, if you want to get up on time every day, put the alarm clock just outside your room so that you have no choice but to get up.

If you can anticipate the situations that may be difficult and have a plan in place to deal with them, you won't have to think on your feet and wrestle with temptation or motivation at a time when you may feel vulnerable.

This is just who I am now – returning to your identity

As motivation rises and falls along the journey of change, returning to your sense of self and the identity you want to create can help you to persist when motivation has disappeared. If you see yourself as someone who looks after their dental hygiene, you pick up the toothbrush every day whether you feel like it or not, because it is just what you do.

Our sense of identity does not have to be entirely fixed by what is laid out for us early in life. We continue to create and build on that identity throughout life with everything we do. When our goals are underlined by our intention to become the person we want to be, or even better, when we have decided that this is who we are now, then we can act in line with that even on the days when motivation is low.

For more on how to steer your sense of identity, see Chapter 33 on how to work out what matters.

 Toolkit: Future self-memory and journalling improves your chance of making better choices
Spend time imagining your future. When we create a vivid image of ourselves in the future, the easier it becomes to make choices in the here and now that will benefit your future (Peters & Buchel, 2010).

Think about yourself at a point in the future and how you might feel about the choices you have made, what you said yes to and what you said no to. How will those choices have affected your life? Which of your choices and actions do you think you will be most proud of? When you reach that time in the future, what will you be focusing on? How will you feel about your past self when you look back?

Pros and cons from Dialectical Behaviour Therapy (DBT)

DBT is a psychological therapy that helps people to find safe ways to manage intense emotions. But the skills taught in DBT can be useful in many aspects of life, including those days when we are trying to stay on track with our goals but we don't feel motivated. Here is one of those skills.

While it is helpful to consider the future that we do want, it is also helpful to look at the future we don't want. In therapy some people spend time exploring in detail the pros and cons of both staying the same and working hard on change. You can use the table below to try for yourself. It pays to spend time being honest with yourself about the true extent of the cost of things remaining as they are. While change inevitably has disadvantages (we may have to tolerate distress and endure discomfort in our efforts), they may be outweighed by the price we pay for staying as we are. This can be a valuable exercise to come back to when we start to feel like giving up on a positive life change or we get off track.

Change

Pros	Cons

Keeping things the same

Pros	Cons

Try this: Building on your identity with intention takes some thought and conscious effort. Try sitting down with pen and paper to write out some answers to these questions. Better still, use a journal and keep coming back to your responses whenever you are working on a life change.

- What is the larger overall change that I am trying to make?
- Why is this change so important to me?
- What kind of person do I want to be as I face this challenge?
- How could I approach this challenge in a way that would make me feel proud when I look back on this period of my life, regardless of the outcome?
- What are those smaller goals that I need to achieve along the way?
- How would I like to face those days when motivation is low?
- Am I listening to my body and what it needs?

Chapter summary

- We cannot rely on motivation to be there all the time.
- We can practise acting in opposition to urges so that we can act in line with our values rather than how we feel right now.
- Repeat a new behaviour enough times and it will become habit.
- For any big goal, rest and replenishment along the way is vital – just ask any elite athlete.
- Make use of small rewards along the way.

CHAPTER 9

Big life changes. Where do I start?

Sometimes you reach a moment in your life when you realize a change is needed and you know exactly what that change is. But it doesn't always happen like that. More often than not, we go through a period of strain and discomfort. We start to recognize that things are not as we would like them to be, but we can't pinpoint exactly why or how we can begin to make it better.

This is where your incredible human brain comes into its own. In chapter 3 we talked about metacognition. Our ability to not only consciously experience the world, but to also then think about and reassess the experience we had. This is a key life skill that we make use of in therapy. It is the epicentre of any big life change. You cannot change what you cannot make sense of.

Albert Einstein reportedly once said, 'If I had an hour to solve a problem, I'd spend fifty-five minutes thinking about the problem and five minutes thinking about solutions.' This quote often comes to mind when I hear the common misconception that therapy consists of sitting in a room and dwelling on your problems. It does involve thinking about your problems, but there is method in that. The most effective way to resolve a problem is to understand the problem inside out.

So how do we use metacognition when we are facing big changes? Building awareness begins with looking back. For anyone in therapy or counselling, you can talk about things that have happened and get useful prompts from the therapist to help you make sense of it. For those who are using a self-help approach, journalling is a great place to start. There is no pressure to write huge amounts or to write in a way that makes sense to anyone else. The aim is to build on your ability to reflect on your experiences and how you responded to them. For example, let's say you failed an exam, and in the moments after finding out, you called yourself a stream of unprintable names and told yourself that you will never amount to anything. Metacognition involves reflecting on those thoughts and how they further impacted on your experience.

The power of using metacognition is that it opens up our ability to be accountable to ourselves and to examine the part we play in staying the same or making change. It reveals the big influence that seemingly small behaviours can have, both positive and negative.

Journalling in this way can feel strange if we are used to glossing over things without paying too much attention to the details. But over time those details can help us to build our awareness of our experience in hindsight, as we start to spot the cycles and patterns of behaviour in the moment, as they happen. This is when we create the possibility of choosing something different and making the positive changes that we want for ourselves.

 Try this: Use these journal prompts to help you explore the problems you are tackling and practise the skill of thinking about your thoughts.

- Describe any significant events that happened.
- What thoughts did you have at the time?
- How did that way of thinking impact on how you felt?
- Describe any emotions you noticed.
- What triggered those emotions?
- What urges did you have?
- How did you respond to the feeling?
- What were the consequences of your response?

Chapter summary

- It's not always clear what we need to change and how to do it.
- You cannot change what you cannot make sense of.
- Get to know your problem inside out to make it easier to identify which way to go next.
- Start by reflecting on situations after they have happened.
- Be ready to get honest with yourself about ways you may contribute to the problem or keep yourself stuck.
- Therapy supports you through this process. But if you don't have access to therapy, journaling can be a good place to start.

3

On
Emotional
Pain

CHAPTER 10

Make it all go away!

If you ever go to therapy, very early on the therapist will ask you what you want from the process. Most people will include emotions in their answer. They have some painful or unpleasant emotions that they want to get rid of, and are missing out on some of the more enjoyable or calm ones that they would like to feel again. And why wouldn't they? We all just want to be happy. They feel at the mercy of those painful emotions and want them gone.

Far from making emotions go away in therapy, you learn to change your relationship with them, to welcome them all, to pay attention to them, to see them for what they are, and to act in ways that will influence them and change the intensity of them.

Emotions are neither your enemy nor your friend. They do not occur because your brain has a few cogs misaligned or because you are a sensitive soul, as you were told in the past. Emotions are your brain's attempt to explain and attach meaning to what is going on in your world and your body. Your brain receives information from your physical senses about the outside world and from your bodily functions, like your heart rate, lungs, hormones and immune function. It then uses memory of these sensations that occurred in the past to make some sense of them now. This is why the heart palpitations brought on by too many cups of coffee can culminate in a panic attack. The pounding heart, faster breathing and sweaty palms all feel a bit familiar to that time you had a panic attack in the supermarket. The physical sensations feel just like fear and your brain gets the message that all is not well, which ramps up your threat response.

Wouldn't it be great if we could wake up in the morning and just decide what to feel that day? Enter love, excitement and joy, please! Unfortunately it is not so simple. The opposite of this idea is that emotions just pop up, with no trigger, and we have no control over what happens or when it happens. All we can do is try to resist them, block them out and be rational. But this is also not the case. While we cannot directly trigger all emotions, we have much more influence over our emotional state than we were ever taught to believe. This does not mean that you are to blame for feeling emotional discomfort. It means that we get to learn about the many ways we can take responsibility for our own wellbeing and construct new emotional experiences.

What *not* to do with emotions

Push them away

Imagine you are at the beach. You walk into the sea up to your chest. The waves need to pass over you to get to the shore. If you try to hold the waves back and prevent them reaching the shore, you learn how powerful those waves are. They push you back and you quickly get engulfed and overwhelmed. But you don't have to tumble and struggle against the waves. Those waves are coming no matter what. When you accept that, you can focus on keeping your head above the water as it passes. You still feel the effects. Might even get lifted off your feet for a moment. But you move with the water and brace yourself ready to land back on your feet.

Dealing with emotion is much the same as standing in the waves. When we try to stop feelings in their tracks, we easily get knocked off our feet and find ourselves in trouble, struggling to catch a breath and work out which way is up. When we allow the emotion to wash over us, it rises, peaks and descends, taking its natural course.

Believe they are facts

Emotions are real and valid, but they are not facts. They are a guess. A perspective that we try on for size. An emotion is the brain's attempt to make sense of the world so that you can meet your needs and survive. Given that what you feel is not a factual statement, neither are thoughts. That is partly why therapies like CBT (Cognitive Behavioural Therapy) can be so helpful to many people. It gives us practice at being able to step back from thoughts and feelings and see them for what they are – just one possible perspective.

If we know that thoughts and feelings are not facts but they are causing us distress, it makes sense to check it out and see whether it is a true reflection of reality or whether an alternative would be more helpful. When we treat our current thoughts and emotions as facts, we allow them to determine our thoughts and actions of the future. Then life becomes a series of emotional reactions rather than informed choices.

So how do we stop buying into thoughts as if they are facts? We ask questions. Something that therapy gives us practice at is being curious about experiences in both our inner world and the world around us. Individuals sit down opposite me and start talking about the things they got wrong that week, and feelings they shouldn't have had, stepping into the old habit of self-criticism and self-loathing. Then we shift our perspective to a bird's-eye-view. We look at how those behaviours fit our formulation. We shift into curiosity, where there is no need for self-attack. So, whether it has been a great week or a hard one, we learn and grow.

Holding curiosity allows us to look at our mistakes and learn, when they might otherwise be too painful to acknowledge. Holding curiosity brings with it a sense of hope and energy for the future. Whatever happens, we're always learning.

Toolkit: Examine your coping strategies

- What are the first signs for you that emotional discomfort is present?
- Is it a behaviour? Do you recognize your blocking or protective behaviours?
- Where do you feel the emotion in your body?
- What thoughts are there? What beliefs are you buying into about this situation? What effect is that having on you?
- Try writing down those thoughts and narratives.
- What can they tell you about what you are afraid of?
- What behaviours tend to follow a strong emotion?
- Do those behaviours help you in the short-term?
- What is their longer-term impact?
- Ask a trusted friend to go over the story with you and help you identify any biases or misunderstandings. Explore with them the different perspectives you might have.

Chapter summary

- Emotions are neither your enemy not your friend.
- We have more influence over our emotional state than we were ever taught to believe.
- Pushing emotion away can cause more problems than allowing it to wash over us and take its natural course.
- Emotions are not facts but are one possible perspective.
- If there is painful emotion, get curious, ask questions. What can they tell you?

CHAPTER 11

What to do with emotions

If you have skipped straight to this chapter, you may be looking for the answer to emotional pain. What Is that thing to make it all go away? Well, if that is you, bear with me. Please don't close the book just yet, but I am about to tell you the opposite of what you're probably hoping to hear.

Somewhere in the middle of clinical training, we had an introduction to mindfulness. You might think that a bunch of trainee clinical psychologists would be open-minded enough to sit and learn patiently. But the room was full of giggles as we all tried to sit in silence and notice what we felt. Clinical training is all about *doing*, getting things done. We were all firmly in 'doing mode'. To shift into just *being* was proving a challenge for everyone in the room, much to the teacher's annoyance. At the time, I admit I was full of judgements about how I couldn't see myself using this or teaching it to anyone.

But it was part of the programme, so I had to try it out. As training progressed and became more stressful, I found myself in the midst of assessment season with a thesis to write and exams looming. Tension was high. One of my favoured tools for stress management at the time was running. I took a break from the desk and went out for a run through the local countryside. My head was buzzing with to-do lists and fears about getting everything done and getting it right. I had another go at using mindfulness, this time while I was moving.

I followed a long gravel track through the woods and listened to the sound of my feet hitting the stones. I allowed the feelings of anxiety and stress to be with me. I didn't try to push them away. I didn't try to plan or problem-solve. Every few seconds my mind would be off, telling me about the things I should be doing instead of this, offering up worst-case scenarios of missing deadlines, failing assignments and an email I needed to send when I got back home. Every time, I let the thoughts come. And every time, I let them pass behind me and returned to the sound of my feet on the gravel. I must have gone through this process a thousand times. Distract, pull back. Distract, pull back. On the way back home, as I approached the end of the track, it hit me. I realized what it was that all the academic texts were struggling to convince me of. I still had all the same hurdles to face. But I wasn't fighting against the tension. I was allowing it to pass – so that is what it did.

The idea of welcoming all emotional experience feels almost alarming at first. It is the opposite of what many of us are taught to do with feelings. We are taught that feelings are the opposite of rationality. Something to be squashed down and hidden, pushed to the back of your mind, unspoken. But allowing them to rise up, and even welcoming it?

Many of us fear emotions – that is, until we begin allowing ourselves to experience them and understand that they all rise and fall like waves.

Mindfulness enables us to use the tool of awareness. Awareness sounds very basic and rather vague but it's the tool we never know we need until we use it. Switching off the autopilot and building awareness of thoughts, emotions, urges and actions adds in an amber light before the green light flashes and we act on an urge or emotion. It offers us the chance to consciously pause where autopilot may have driven us forward. So we give ourselves more chance to make different choices based on our values rather than simply responding to emotion.

As an artist works closely on a small detail of a large painting, she will occasionally take a step back and check that each new action fits in with the vision she has for the whole picture. The metacognition tool of pausing between emotion and action is that same process of stepping back, even just for a moment, to check in with your thoughts and actions to see if they will be in line with the person you want to be. The ability to check in on the bigger picture, even for the smallest of moments, can have a powerful impact on the way we live life.

As the river of constant thoughts flows, we get to put our head above water and check those thoughts are heading in the direction we want to move in. We can consider that direction in the context of our meaning and purpose, and not just move with the flow because that is the way it naturally goes.

See emotions for what they are

Seeing emotions for what they are is key to being able to process them in a healthy way. You are not your feelings and your feelings are not who you are. The sensation of an emotion is an experience that moves through you. Each emotion can offer you information, but not necessarily the whole story. If there is something emotions are pretty useful for, it's telling you what you need. When we allow ourselves to feel emotion without blocking it out or pushing it away, we can turn towards it with curiosity, and learn.

Discovering what we need is even more valuable if we then use that information to do what is necessary and meet those needs. I think it is always useful to start with the physical. As we discussed in previous chapters, no amount of therapy or psychological skills is going to overturn the destructive impact of poor sleep or diet and lack of physical activity. Once we start taking care of the body we live in, we are already well on the way to being able to work on the rest.

Pick a name

When you feel something, give it a name. Learn lots of names for lots of emotions. We don't only feel happy, sad, fearful or angry. We feel vulnerable and ashamed, bitter and grateful, inadequate and excited.

In therapy a lot of work is put into this. Notice what you feel, notice where you feel it in your body, and label it. It is common for people to recognize the physical sensations, but have no idea what that emotion is; perhaps a legacy of past teachings that we don't talk about emotions. Nobody needed a name for each different emotion because they never voiced it out loud. But they were able to pick out the physical manifestation of it because it has always been more acceptable to tell someone you feel sick and have a pounding heart, than to say you feel vulnerable and insecure.

Increasing your emotional vocabulary so that you can distinguish finely between different emotions helps you to regulate those emotions and choose the most helpful responses in social situations (Kashdan et al., 2015).

Self-soothing

When painful emotions become intense, it's very easy to say they will rise, peak and descend, but the reality of that experience can be excruciating, and lead to strong urges to do unhealthy or even dangerous things to make it all disappear sooner.

While some self-help books might tell you that you can just think positive thoughts to change how you feel, I would say that's going to be a struggle. Trying to change how you think is hard enough when you are feeling fine. Trying to change the thoughts that arrive at the height of distress feels nearly impossible. When we are overwhelmed, the best strategy is to step back and be mindful of that emotion as much as possible, see it as a temporary experience, and to turn down the dial on our threat response by soothing our way through it.

In a therapy called Dialectical Behaviour Therapy (DBT) we teach individuals how to soothe their way through distressing emotions with simple skills that help you to ride the wave until it comes back down. These are called distress tolerance skills. One of those skills is called self-soothing (Linehan, 1993).

Self-soothing is any set of behaviours that help you to feel safe and soothed as you experience a painful emotion. When your threat response is triggered, the message being received by your brain is, 'We are not safe! All is not OK! Do something about this now!' If we want that distressing emotion to stop escalating and start its process of coming back down to base-line, we need to feed our body and brain new information that we are safe. There are lots of ways we can do this, because your brain takes its information from each of your senses. That means you can use each of those to send messages to your brain that you are safe. Your brain also takes its information from the physical state of your body, including your heart rate, breathing rate and muscle tension. This is why physical experiences that relax muscles, such as a warm bath, can be effective in contributing to getting you through distress.

Other self-soothing ideas include:

- A warm drink
- A chat with a trusted friend or loved one
- Physical movement
- Calming music

- Beautiful images
- Slow breathing
- Relaxation techniques
- A scent or perfume you associate with safeness and comfort

One of the quickest ways to tell your brain that you are safe is actually through your sense of smell. Finding a scent that you associate with safety or comfort, maybe the perfume of a loved one or a lavender scent that you find calming, can be helpful in helping you to focus the mind and calm the body at the same time. For those who struggle with distressing emotions when out in public, here's an example. A popular choice in therapy has been to carefully unstitch a soft toy, fill it with lavender, and sew it back up. Then whenever you are out in public and start to feel overwhelmed, you can breathe in the scent and ground yourself, soothe yourself through it without anyone even noticing.

A great tool that is often used in DBT is to create self-soothing boxes. The reason this is such a great idea is that when you are in emotional pain, at the height of your distress, your brain is set up to bypass your problem-solving capabilities. If you are under threat, you don't have time to think things through. This is when your brain makes a quick guess for you and acts on impulse. A self-soothing box is something you prepare in advance, when you are able to think through what most helps in times of distress. Grab an old shoe box and fill it with anything that could help to soothe you when you are in distress. As described above, anything that you associate with feelings of safety and comfort is great to include. I have a self-soothing box in my therapy room that I use as an example. Inside that box is a note to call a particular friend. Seeking help may not be our first thought when we are struggling, but following a simple instruction to call a trusted friend can steer us in the right direction. As we know from previous chapters, human connection helps us to recover from stress more quickly. Other things that I included in my self-soothing box are a pen and pad of paper. If you cannot bring yourself to talk, then expressive writing has been shown to help us to process emotion and make sense of what is happening.

The box could also include lavender oil (or any scent that you associate with comfort), some photographs of people who care about you and whom you care about, and a list of soothing or uplifting music tracks to play. Carefully chosen music can have a powerful impact on our emotional state. Create a playlist of tracks that will help you to feel calm, safe and soothed when you are distressed.

My box also includes a teabag, because in England we associate tea with comfort and connection. Including something like this in the box is simply a clear instruction to follow when you might otherwise be struggling to think through what you need.

Crucially, keep that box somewhere that is easy to find when you might need it. This tool is all about making things easier for you to cope in the way that you want to during the toughest of times, to help you stay away from the less than healthy habits when you are vulnerable to using them.

Chapter summary

- You are not your feelings and your feelings are not who you are.
- The sensation of emotion is an experience that moves through you.
- Each emotion can offer you information but not necessarily the whole story.
- If there is something emotions are useful for, it's telling us what we need.
- When you feel something, give it a name. Try to label emotions with more detail than just happy or sad.
- Allow emotions to be present and soothe your way through, rather than blocking them out.

CHAPTER 12

How to harness the power of your words

The language we use can have powerful effects on our experience of the world. It is our tool for making sense of things, helping us to categorize sensations, learn from past experiences, share that knowledge and predict and plan for future experiences.

Some words for emotions have gradually been used more and more to mean different things, until they have taken on broad and vague meanings. 'Happy' has become an umbrella term for anything positive, to the extent that nobody really knows if what they are feeling qualifies as 'happy'. If I feel passionate, am I happy? If I feel calm and content, am I happy? If I feel inspired and energized, is that happy?

The same has happened to words like depression. What is

a depressed mood, exactly? Sadness? Emptiness? Agitation? Numbness? Uneasiness? Restlessness? Flatness?

And does any of this really matter? Well, as it turns out, it does.

Having fewer concepts or words to differentiate discrete negative emotions is associated with higher levels of depression after stressful life events (Starr et al., 2020). Those who are able to distinguish between negative feelings tend to be more flexible in how they respond to problems. For example, they are less likely to binge drink under stress, are less reactive to rejection and show less anxiety and fewer depressive disorders (Kashdan et al., 2015). This does not mean that those things are caused by the difficulties in distinguishing between negative feelings, but it does show that we have a powerful tool that we can use to help us through difficult times.

The more new words you can build up to differentiate between feelings, the more options your brain has for making sense of various sensations and emotions. When you have a more accurate word for a feeling, this helps to regulate your emotions and in turn means less stress for your body and mind overall. This is a crucial tool if you want to be more flexible and effective in how you respond to challenges that you face (Feldman Barrett, 2017).

The good news is that this is a skill we can all keep building upon. Here are some ideas for how to build up your emotional vocabulary.

- Get specific. When you feel something, try to go beyond 'I feel awesome,' or 'I'm not happy.' What other words can you use to describe this feeling? Is it a combination of feelings? What physical sensations do you notice?

- One emotion label may not be enough to encapsulate this feeling. Is it a combination of feelings? For example, 'I feel nervous and also excited.'
- There is no right or wrong way to label an emotion. The key is to find a description that you and the people around you can become familiar with. If you can't find the words, you can make up your own or find words from other languages that don't have a clear translation.
- Explore new experiences and play around with ways to describe those experiences. From tasting new foods, to meeting new people, reading books or visiting somewhere new. Each new experience offers the opportunity to view things from a different perspective.
- When it comes to building up the ability to describe those new experiences, take every opportunity to learn new words. This does not have to only come from books (although it can). It can also come from music, movies and any other place that exposes you to new words to describe how you feel.
- Write down experiences and explore ways to describe how you felt. If you often feel at a loss for words when trying to describe how you feel, and need help to build up your emotional vocabulary, the Feeling Wheel (Willcox, 1982) is a great resource that is often used in therapy for exactly this purpose. You can keep a copy in

the cover of your journal and use it to find more specific words. You can also use the blank spaces to add in your own words as you come across them in other places.

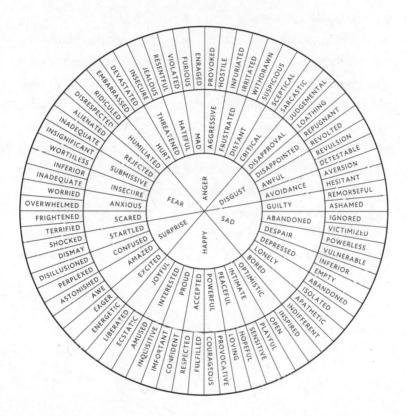

Figure 7: Use the Feeling Wheel (Wilcox, 1982) to help you find the words to describe how you feel.

117

Don't only focus on the negative emotions

Journalling is a helpful way to process and understand the experiences and emotions that happen to you. But journalling is not only helpful for making sense of difficult experiences. It's also important to spend time writing about positive experiences, even small moments that felt positive. This is because every action is a specific pattern of neural activity in the brain. When you repeat an activity over and over, that neural pathway is strengthened and becomes easier for your brain to access. So if you want to be able to more easily cultivate positive feelings, thoughts and memories, put in the practice by keeping a journal. When you get practise at cultivating certain feelings and experiences in this way, it gets easier to access those feelings in the future.

Chapter summary

- The language we use has a powerful effect on our experience of the world.
- The more words you have to describe how you feel, the better.
- If you don't have the words you can use something like the Feeling Wheel to give you prompts.
- Notice the words others use, read books and explore the ways you can keep building your emotional vocabulary.

CHAPTER 13

How to support someone

If you are supporting a loved one as they struggle with their mental health, it is quite possible that you feel completely inadequate. You don't know how to fix it, you're not sure what is the right thing to say. You want to make it all OK for them. But you can't. So you feel lost, desperately wanting to help and not knowing how.

When someone we love is suffering, sometimes the stress that it brings up can give us the urge to escape how we feel about their pain. But when we do that, we can feel even more helpless and paralyzed because we stop ourselves from providing even some low-level support that could help us to feel more confident in the supporting role (Inagaki et al., 2012).

While there are no hard and fast rules for supporting someone through mental health problems, there are a few things that can help along the way.

1. When we focus on trying to fix the problem, it is easy to underestimate the power of simply being there. Most people don't want to be told what to do. But they do want someone to keep showing up to check in and show they care.
2. If the loved one has a specific diagnosis, it can help to learn about how it affects them and get more specific advice on the challenges they are facing.
3. Don't forget that the person you are supporting will have an idea of what they need. So asking them how they would like you to support them can help to give you guidance, while also communicating to them that you are listening.
4. Caring for someone can put a strain on your own mental health. But you cannot support them at your best If your mental health starts to deteriorate. So it is absolutely imperative that you prioritize your own health too – even in small ways. Keep a close eye on the basics. Keep track of your sleep, routine, nutritional intake, exercise and social contact.
5. Get your own support. Whether it be someone you trust, a support group or a professional, having a safe space to talk about your own feelings and think about how to move forward can help to stop you burning out.
6. Set boundaries. Supporting someone else does not mean that your life no longer matters. Getting clarity on your own values can help to keep you going when things

get tough, but also to ensure that you can maintain some balance.

7. Work on a crisis plan. If the person you are caring for ever feels unsafe, then it is important to have a crisis plan. It doesn't need to be complex. Acknowledge any early warning signs that things may be deteriorating and list the things that both of you can do to ensure everybody's safety in that scenario. Having a written plan with all the numbers you need to call makes it easier to do what you need to in a crisis.

8. It is easy to underestimate the power of listening with compassion, kindness and curiosity. The problems may not disappear when you do that, but you are helping that person to feel cared for and less alone, which greatly improves their chances of recovery. Social support is a powerful tool and it does not need to come with all the answers, just a big dose of compassion.

9. Supporting someone does not mean that you have to connect with big, intense conversations. Human connection in the smallest of moments matters. Walking while you are talking can help those who feel uncomfortable with opening up. You can also say nothing at all. Simply spending time together, even in silence, is OK. By being there, you are helping them to feel less alone and more cared for.

10. If you are trying to help that person open up about their struggles, using open questions that ask for more than a

yes or no answer can help. For example, rather than 'Are you OK?' try 'What are you thinking?'

11. Listen carefully. Don't offer advice unless they ask for it. Just reflect back what you hear them saying; let them know they are being heard and respected.

12. If that person talks about feeling hopeless or helpless, says that they cannot see a way out, or you become concerned about their safety, always seek professional advice.

13. Don't underestimate the power of practical help. If someone is facing struggles with mental health, physical health, pre- and post-natal periods, or grief, all of these can make the usual day-to-day tasks more difficult. For example, helping someone to eat a few healthy meals each week by turning up with a home-made dinner is a great way to support a loved one.

14. Being sensitive to situations in which your loved one may feel especially vulnerable (or asking them if you don't know) means that you can be there for them when they most need it. For example, if a loved one is recently bereaved and has to attend a social function alone for the first time, don't avoid them. Lean in and show them love and kindness. Those situations will still be hard, but feeling less alone can mean everything.

15. It's OK to change the subject. Being around someone does not mean you have to focus on their struggles the whole time. Distraction can be a welcome relief that they might find difficult to achieve when alone.

16. Have no expectations about their healing or recovery. It is never smooth and linear. There will be good days and bad days. Being surrounded by loved ones who are accepting of those ups and downs over the years will help them to do the same.

17. Be honest. If you want to be supportive but you are not sure how, say that out loud. Ask the person to let you know if you are saying or doing something that is not helpful. This openness allows everyone to feel less anxious and to truly connect, knowing that the situation is working for everyone.

Chapter summary

- It is normal to feel overwhelmed or inadequate when supporting someone with mental health problems. You want to fix it but you don't know how.
- Leaning in to support someone who is suffering can be stressful as you don't want to say the wrong thing. But don't avoid them.
- You don't have to fix everything to be a great support.
- Look after yourself to prevent burnout. Get your own support and set clear boundaries.
- Never underestimate the power of listening.

4

On Grief

CHAPTER 14

Understanding grief

We often associate grief with the death of a loved one. But we can grieve at other times too. Endings that feel significant to us can trigger a grief reaction – even if the ending was not caused by death.

We have fought our way through a pandemic that changed our lives. Along the way we lost friends, family, livelihoods, jobs, family businesses built over generations. We lost financial security, final moments with loved ones, and precious time to hold and be close to them. We lost a sense of certainty about what the future holds, and access to the social support that would help us cope with that. The profound losses experienced by so many have changed the world and left a psychological fallout that is heavy with grief.

For anyone feeling the effects of a loss, these are some things to remember.

Grief is normal

I have met so many people who tell themselves that they are failing at life because they are struggling to cope with grief. They make broad judgements about their strength of character, as if grief is a disorder or a problem they should have fixed. Grief is a normal part of human experience. It is a necessary process to go through when we experience the loss of someone or something that we loved, needed, felt connected to and that held meaning in our life.

Sadness can be a part of grief. But there is much more to grief than sadness. It can be a deep yearning for the person who is gone. Relationships are at the core of what it means to be human. Among the people I have met in my career so far, their human connections have been the most meaningful aspects of their lives. When the relationship ends, the need for that connection doesn't stop.

Your body grieves too. As explained in previous chapters, everything we think and feel happens within the body. Grief is no different. The loss of a loved one is a huge psychological and physical threat. The pain can feel both emotional and physical. The stress response is repeatedly triggered.

In talking about things that help us through grief, let's be clear on what help means. Things that help do not make the pain disappear or make us forget or force us to let go. Help might be as simple as finding out that the rollercoaster of emotions you feel is normal. It might be finding new ways to sit with and process the pain in a safe and healthy way.

Grief can feel intolerable. It makes absolute sense that our most natural human response to that might be to block it out. The pain is so intense and vast that it is terrifying. So we push it away if we can. But when we block one emotion, we tend to block them all. We can be left feeling hollow, numb and struggling to find meaning and to engage with life in the way we once did.

If we find a way to push it all down under the surface, maybe by keeping really busy, numbing ourselves with alcohol or with denial of what has happened, we may feel like we're doing OK. Then something small that seems insignificant blows the lid off and this world of pain explodes, leaving us in shock and questioning whether we can cope.

Unresolved grief is associated with depression, suicidality and alcohol abuse (Zisook & Lyons, 1990). So denying our grief and pushing it away feels like self-protection but in the longer term can be the opposite.

How easy it is to say all this. How hard it is in reality to experience it. When we block out pain, it is for good reason. An ocean of grief as deep as it is wide, it feels too big, too much, never-ending. How do we possibly face something like that? We can start by understanding what to expect. We can also make sure we get to know what helps us navigate the experience. Then we take things one experience at a time. We take a few strides into that ocean of grief. We feel it. We breathe. We step back and rest for a moment. Over time we learn to take more steps, go deeper, immerse ourselves, knowing that we can safely return to the shore. Feeling the grief does not make it disappear. But we build up our strength to know that we can be reminded and yet still return to engage with life as it is today.

Chapter summary

- Endings that feel significant can trigger a grief reaction – even if the end was not caused by death.
- Grief is a normal and natural part of being human.
- The pain can feel both emotional and physical.
- Things that help do not make the pain disappear or force you to let go.
- Trying to completely block out grief can lead to problems further down the line.

CHAPTER 15

The stages of grief

You might have heard about the stages of grief as originally described by Elisabeth Kubler-Ross (1969). Since then, it has been established that they are not experienced in stages and they do not happen in any particular order or time-frame. But they do describe some of the most common experiences that can be a part of normal, healthy grief. It is important to remember that these are not a prescription for how you *should* be grieving. They are not a rule book for the best type of grieving. They are descriptions of experiences you might notice along the way. So if you recognize any of them in your own experience, or in a loved one, you know these are a part of normal, healthy grief.

- Denial

  Denial and shock can help us survive the overwhelming pain of grief. It does not mean that we deny any of it is happening. But there may be a gradual pacing of how you take on board the situation you are facing and the new reality that awaits you, whether you chose it or not. Over time, denial begins to fade and this allows new waves of emotion to surface.

- ## Anger

Underneath anger is often intense pain or fear. When we allow ourselves to truly feel that anger and express it, we can bring those other emotions to the surface and work on them. But many people have been taught to fear anger and to feel ashamed of expressing it. So we hold it under the surface, but like holding air under water, it soon bubbles up in another time or place. An outburst at a friend or doctor or family member that appears out of character.

Anger is there to agitate us to move and make something happen. When we experience anger about something that we cannot control, using physical movement helps us to use that physiological arousal in the way it was designed to be used. Venting in that way can be helpful to use up that energy generated by anger and bring ourselves back down to baseline, at least for a while. Once the body is calm, you are then more able to access the cognitive function necessary to get clarity on your thoughts and feelings or any problem-solving to be done. It can be helpful to do that with a trusted friend or loved one who can support you, or by writing things down. We know from the research that ruminating alone on angry feelings can make the anger and aggression more intense rather than less (Bushman, 2002).

Trying to do any sort of deep relaxation exercise before you have used physical action to move through the anger and bring your level of arousal back down may be too difficult. But once you have expressed it in any way that best suits you, using guided relaxations can be helpful to replenish the body and mind until the next wave of anger arrives.

• Bargaining

Maybe this occurs in fleeting moments. Maybe it's hours or days spent ruminating over the 'What if . . .' and 'If only . . .' thoughts. This can easily lead down a path towards self-blame. We start to wonder what could have been different if we had made different choices at different times. We may start bargaining with a god, if we have one, or the universe. Or maybe we promise to do things differently from now on and devote our life to making things better in some way, trying desperately in our mind to make it all OK again. We just want things back the way they were.

• Depression

The word depression here is used to describe that deep loss, an intense sadness and emptiness, that follows after a bereavement. This is a normal reaction to loss and does not necessarily indicate a mental illness. Depression is a normal response to a depressing situation. Sometimes people around us can feel frightened by it and naturally want to fix it or cure it, or worse, want you to snap out of it.

But recognizing depression as a normal part of healthy grief means that we can try to soothe ourselves through that pain and work hard to re-engage with normal life and look after our wellbeing as much as we can. The ideas and tools covered in Section 1 still apply here. But we also don't have to deny pain, or push it down and hide it, as I will explain later.

• Acceptance

When we give grief the time and space it needs, we begin to feel more able to step forward and play an active part in life again. Acceptance can be misunderstood as agreeing with or liking the situation. That is not true. In acceptance, the new reality is still not OK. It is still not as we want it to be. But we begin to take on the new reality, listen to our needs, open up to new experiences and make connections.

It is also important to point out that acceptance is not an end point in grief. It might be fleeting moments in which you have found a way to live in this new reality. There may be other moments when you return to bargaining and yearning for that person. Going back and forth between these states is normal, and it's to be expected as you face all the new challenges and experiences in your life. This means if you have started to find new moments of contentment or joy, things seem to be going OK, and then you find yourself overwhelmed with a wave of anger or sadness (or anything else), that does not mean you have gone backwards. You are not getting grief 'wrong'. Grief comes in waves that we can't always predict.

Chapter summary

- Denial can help us survive the overwhelming pain of grief. As denial fades, this allows new waves of emotion to surface.
- When we experience anger about something we can't control, using physical movement helps us to use the physiological arousal and bring the body back down to calm for a while.
- Ruminating over the *What ifs* can easily lead down a path of self-blame.
- Depression is a normal reaction after a bereavement.
- Acceptance is not the same as liking or agreeing with the situation.

CHAPTER 16

The tasks of mourning

So how do we begin to get through this intense, confusing and often chaotic experience that we call grief?

William Worden (2011) described what he thought were the four tasks of mourning.

1. To find some acceptance in the new reality after the loss.
2. To work through the pain of grief.
3. To adjust to an environment in which that loved one is missing.
4. To find a way to keep a connection with them in a new way while also engaging in life as it is now.

After a loss, people deal with their grief in different ways. While some are oriented towards feeling the pain and emotions that arise, others focus on trying their best to distract themselves from the overwhelming emotions. Neither of these are wrong. In fact, we need both. We cannot work through grief all in one go and feel that much emotional pain without rest. But we cannot do the work through grief without allowing ourselves the space to feel it. So the work becomes a process of movement between feeling the pain and replenishing the body and mind with something distracting or comforting that allows you a break between the waves of emotion (Stroebe & Schut, 1999).

So spending time with the emotions that come up (whether you made an active choice to go there, by looking in a memory box or visiting a memorial, or if that emotion rose up without choice) is a necessary part of that process. It allows the feelings to unfold and be expressed, through talking, writing, or weeping. When you feel you need to step back from that, it helps to turn your attention to something that brings the stress response back down. Using the self-soothing skills from Section 3 can be helpful (see page 109), especially when the pain has been overwhelming. Grounding techniques may also be helpful here. But there is no set prescription, as every individual, every relationship and therefore every grieving process is unique. The key is to find something safe that allows you some time for restoration, even if that time is short.

One of the problems with the 'trying to get on with it' mode and not allowing ourselves to focus on the loss at any point is that it can take relentless effort, and there is no break from that. We might then need to stay busy for fear of being overwhelmed if we press pause. So we become stuck. We're not able to rest because the work of keeping the pain at arm's length is constant. When pain is vast, the actions we take to push it down and keep it under the surface can cause damage to both the individual and their relationships. If you disconnect with one emotion, you disconnect from them all.

Feel whatever is there

It is OK, when you are grieving, to feel everything. It is OK to feel despair. It is OK to feel rage. It is OK to feel confused. It is also OK to feel joy. It is OK to smile if that is where the moment takes you. It is OK to enjoy the warm sun on your face for a moment, or laugh at someone's joke. It is all OK. It is normal to feel guilty when you start to allow yourself to live again, but allowing small moments of joy to wash over you is just as important to the process of grieving as allowing the pain to. Over time we learn to engage with life and recognize that does not mean forgetting. The love and connection continues.

Small steps forward every day

Do not underestimate the power of the smallest steps forward. If standing upright and washing your face every day feels like a battle, then let washing your face each morning become the current goal. Meet each chapter from where you are and push it where it moves.

No expectations

Expectations about how you should feel, how you should behave, and how quickly you should heal, only make grief harder. Many of those expectations come from a history of our misunderstanding about grief as a taboo subject. Thanks to some pioneers in this area of research, we now have a much better understanding of the process of grief and how to help ourselves through it. The expectations lead people to falsely believe that they are going mad, getting it all wrong, weak and alone. In reality, all the feelings, the ups and downs, are all a normal part of the process. The lack of conversation about grief leads us to worry about whether we are getting it right. The opposite to this, and a much more helpful approach, is cultivating a compassionate connection with yourself and others. One that allows you to express your feelings in a safe place.

Expression

Expressing how you feel is not always easy. Some have the urge to talk. Others clam up and can't find the words. If you want to talk, find someone whom you trust and start talking. If you have all of those very normal fears of being a burden, upsetting the other person – say so. A good friend will tell you what they can manage.

If you cannot talk, write. In whatever way the words come to you. The act of getting those thoughts and feelings out on to the page can help to unravel some of what is going on in your mind and body. It is through the processing of those painful feelings that the work of grieving is done.

Some people find expression through painting, music, movement or poetry. Whatever offers a safe avenue for you to release and express that raw emotion is worth making time and space for. If you are not sure where to start, just start with anything that comes naturally to you. Start with the thing that has helped in the past. Or start with something just because you are curious about how it might be.

If you don't have a therapist to hold the boundaries for you, this is something you can do to ensure that you step into and back out of the emotion. There is a time to feel and a time to block, a time for turning towards and a time for turning away to rest your mind and body. So if you are going to spend some time releasing and expressing that emotion, have those safety nets that help you back.

Remember and keep living

When remembering someone causes pain, and engaging in the present without them causes pain, the two experiences can feel at war with each other. The demands of life keep coming and just one memory can crop up and bring you to your knees.

Perhaps one of the things that changes over time with grief

is the bringing together of these two things. Or the discovery, through trying, of a way the two needs can co-exist. The need to engage in life and the need to remember and stay connected with the person lost. This could be making time for moments that celebrate their life, spending time in rituals that help you to continue your relationship with them, while also making deliberate choices every day to live in a way that honours both past and future.

The work of grief appears to be about stepping into your pain, allowing it to wash over you, soothing and supporting yourself through it, and stepping out of it again and into life as it is now, finding ways to rest and nourish your body and mind through the exhaustion of grief (Samuel, 2017).

Grow around the wound

The wound that is left following a loss is not something to be fixed or healed. We do not want to forget that person, we want to remember them and continue to feel connected to them. So the wound does not diminish or disappear. It remains while we work hard to build a life around it (Rando, 1993). This is a concept that many people find helpful in therapy. The person is just as important to you as they ever were and so the pain of losing them continues. But we find a way to acknowledge their life while beginning to grow and create a life with meaning and purpose alongside that grief.

You find ways to remember, celebrate and feel your

connection to that person, and to keep living. You learn that pain and joy, despair and meaning can all be a part of life. You learn what you are capable of surviving, the depths from which you can pull yourself up, and from that, you keep going.

When to get professional help

Going to a counsellor or therapist does not mean you are getting the grief thing wrong. We need support to get us through the pain of grief, but not everybody has someone that they trust or want to talk openly with. The therapy room can become a sanctuary. A safe space to release raw emotion with someone who is trained to sit firm with you through that. The therapist can help you to make sense of things, use skills to help you manage safely, understand more about grief, and listen in a way you have never been listened to before, without judgement, advice or attempts to minimize and fix things for you. A therapist knows that the work of grief is through the pain, and their work is to walk through it with you and offer a guide when you need it.

Chapter summary

- Grief demands that we work through the pain.
- It takes time to adjust to a life in which the loved one is missing.
- We need to find a way to keep that connection going with the loved one without their physical presence.
- Acceptance of the new reality means we can continue to engage with the things that matter to us. Whatever you feel, it's ok to feel that.
- Do not underestimate small steps and steady progress.

CHAPTER 17

The pillars of strength

The grief psychotherapist Julia Samuel set out the key structures that support us to rebuild our lives through grief (2017). She calls them 'pillars of strength' because they take work and persistence to build. By nurturing each of them, we gain a stable structure to help us through. The pillars of strength are listed below.

1. **Relationship with the person who has died**
   When we lose a loved one, our relationship and love for that person does not end. Adapting to the loss involves finding new ways to feel close to the loved one. For example, visiting a special place that you shared or spending time at a grave or memorial.

2. **Relationship with the self**
   Every other section in this book touches on self-awareness, and working through grief demands the same.

In understanding our own coping mechanisms, finding ways to get support and look after our health and wellbeing throughout, we must listen to our own needs as best we can along the way.

3. Expressing grief

There is no correct way to express grief. If you prefer to do that through quiet reflection, memorials or sharing with friends, the act of allowing yourself to feel whatever comes up and expressing that helps the natural process. When emotions are especially overwhelming, you can make use of the skills offered in Section 3 to help (see page 99).

4. Time

Putting an expectation on how much time you should need for grief is setting yourself up for struggle. When everything is overwhelming, it helps to focus only on each day as it comes until you feel strong enough to take a broader view of the future. Adding pressure to feel a certain way in a particular time-frame only adds pain and distress.

5. Mind and body

As I covered in Section 1, our physical state, emotions, thoughts and actions are like weaves in a basket (see page 53). We cannot change one without influencing the other. This makes taking care of all of those aspects of our experience even more important. Regular exercise, eating well and ensuring we maintain some social contact will all help us to strengthen our mental health when we most need it.

6. Limits

When loved ones around us may be full of advice about how we should be managing and when we should be getting back into everyday life, remembering our capacity for holding boundaries becomes an essential tool. If we are building self-awareness and listening to our own needs, sometimes we need to put boundaries in place and maintain them in order to do what is in our best interests.

7. Structure

I have spoken in previous chapters about our human need for balance between predictability and adventure, structure and flexibility. When our mental health becomes vulnerable after a loss, it makes sense to offer a degree of flexibility that allows for grieving, while also maintaining some level of structure and routine that helps prevent deterioration of your mental health from the absence of healthy behaviours such as exercise and social contact.

8. Focusing

When there are not enough words to describe the sensations that we feel, focusing our attention on simply observing our internal world and visualizing those sensations in the body can help to build awareness of our own shifts in emotional and physical state.

Chapter summary

- We can rebuild a life after a bereavement with time, work and persistence.
- Create new ways you can feel close to your loved one with a special place or memorial.
- Listen to your needs as much as you can along the way.
- There is no correct way to express grief.
- Drop any expectations about how much time you should spend grieving.

5

On Self-doubt

CHAPTER 18

Dealing with criticism and disapproval

Criticism and disapproval is something we all have to face at some point. But nobody ever really teaches us how to deal with it in a way that allows that feedback to enhance our life instead of destroying our self-esteem.

Even the anticipation of criticism or disapproval can be enough to cause a paralysis in our ability to strive for things that matter most to us. So not having the skills to deal with criticism or disapproval in a healthy way can cost you.

Now, this chapter is not about to tell you to just stop caring what anyone else thinks of you. In fact, we are built to care about how we are being perceived by those around us. Criticism can be a sign that we haven't lived up to expectation in some way and sometimes (but not always) can signal a risk of rejection or abandonment. So receiving criticism will naturally trigger your stress response. That response gears us up, ready to do something about it. Historically, rejection from our community was a serious threat to our survival. These days, things are different in some ways, but similar in others. Rejection and loneliness continue to be a big threat to our health and the brain continues doing its job of trying to keep us safe in a group.

Beyond simply keeping us safe, our capacity to imagine what others might think of us is a key skill that helps us to function in the social groups that we live in. We develop our sense of self and identity, not only from our own experience and how we interact with others, but also through what we imagine those other people really think of us, the ideas and perceptions they might have of us. This is called the 'looking glass self' (Cooley, 1902). So it makes sense that what I believe you think of me is going to influence what I do next.

So when we try to tell ourselves to just stop caring what anyone else thinks, even if we feel a momentary boost, the impact of that is often short-lived at best.

People-pleasing

People-pleasing is more than just being nice to people. Anyone would recommend being nice to people. But people-pleasing is a pattern of behaviour in which you consistently put all others before yourself even to the detriment of your own health and wellbeing. It can leave us feeling unable to express our needs, likes and dislikes, and unable to hold boundaries or even keep ourselves safe. We say yes, when actually we want and need to say no. We feel resentful of being taken advantage of, but unable to change it by asking for anything different. And the fear of disapproval never disappears because there is always the possibility of putting a foot wrong, making a wrong choice and displeasing someone – even if that person is someone we don't like or spend time with.

While it is in us all to care about the approval of our peers, people-pleasing takes it much further. If we grow up in an environment in which it is not safe to disagree or express difference, if disapproval is expressed with rage or contempt, then as children we learn how to survive that environment. Keeping other people happy becomes a survival skill that we hone and perfect throughout childhood. It is only later, as adults, that those behaviour patterns become detrimental to our relationships. We second-guess every move we make, always tentatively trying to work out what others are expecting of us. It may even prevent us from making new connections as we hold back on interactions when there is no guarantee that the other person likes us back.

Living a life of people-pleasing is further complicated by the fact that other people don't always voice their disapproval with criticism. We can fear and feel disapproval even when the other person never says a word. When we don't have that information, our mind starts to fill in the blanks for us. The spotlight effect is a term originally coined by Thomas Gilovich and Kenneth Savitsky (2000) to describe the tendency of humans to overestimate how much others are focused on us. We are each at the centre of our own spotlight of attention and we tend to imagine that others are focused on us too, when in reality, everyone's spotlights are usually on themselves. So we can often make the assumption that others are judging us negatively or disapproving when they may not be thinking about us at all.

Those who feel socially anxious tend to focus their attention more on how they are being perceived by those around them (Clark & Wells, 1995). But those who feel more confident tend to have a more outward focus of attention, leading with a curiosity for other people.

So, if we have this brain that is set up to care a great deal what everyone else is thinking, or maybe we notice a tendency towards people-pleasing patterns, how do we live alongside that? How do we ensure that we can have those meaningful relationships but not become trapped by constant worries about disapproval and judgement? And how can we pick ourselves back up when disapproval from someone else stops us from living in line with what matters to us?

The tasks of dealing with criticism:

- Building up the ability to tolerate the criticism that could be helpful and use it to your advantage while maintaining a sense of self-worth.
- Being open to learning from negative feedback that could help you make progress.

- Learning to let go of criticisms that reflect the values of someone else rather than your own.
- Getting clarity on which opinions matter the most to you and why, so that it becomes easier to know when to reflect and learn and when to let go and move on.

Understanding people

Most people who are highly critical of others tend to be highly critical of themselves also. It can reflect the way they have learned to speak to themselves and everyone else. They criticize because it is what they do, not necessarily because it is any reflection on your worthiness as a human being, especially when it is a very personal attack on your character as opposed to anything that could be in any way helpful to you on your path.

As humans we also have a tendency towards egocentric thinking, which can play out in our insistence that other people live by the same values and obey the same rules that we have set for ourselves. This means that criticism can often be based on the critical person's view of the world, neglecting the fact that we all have different life experiences, values and personalities.

Understanding that people tend to criticize others based on their own rules for living is helpful to remember, especially for anyone with that tendency towards people-pleasing. We want everyone's approval, but if each person is unique with their own ideas and views, then we simply cannot please everyone all of the time. If we have a close relationship with the person, then we are likely to value their opinions more (which can make the criticism more painful), but we also may have the insight to better understand what lies behind the disapproval.

Context is everything, but we don't always have access to it. When we don't have that context, it is much harder to see the criticism for what it is – one person's idea that is wrapped up in their own experiences. The natural instinct is to take on the criticism as a factual statement that says something about who we are and to start questioning our own self-worth.

Nurturing self-worth

Not all criticism is bad. When feedback focuses on a specific behaviour, we tend to feel guilty, which prompts us to correct our mistakes to repair the relationship. When criticism attacks our personality and our sense of worth as a person, we tend to feel that in the form of shame.

Shame is the intensely painful feeling that can be mixed with other emotions like anger or disgust. It is different from embarrassment, which is less intense and tends to be felt in public. Shame is much more painful. We feel unable to speak, think clearly or do anything. We want to disappear and hide. The sheer intensity of the physical reaction makes it difficult to recover from.

Shame triggers our threat system in such a way that it can feel like someone held a match to all our other emotions. So we get a rush of anger, fear or disgust to go along with it. Then the self-attack comes swarming in like soldiers over a hill, coming at you with self-criticism, self-denigration and blame. With that sort of onslaught the instinct is to block it all out. But shame is not easy to ignore. So we go for the most absorbing, addictive behaviours that offer instant relief.

Shame resilience is something we can learn, but it is a life practice. Building resilience to shame does not mean you never feel it. Instead, it means we learn how to dust ourselves off and get back up.

Being able to experience shame and come back from it without losing your sense of self-worth involves:

- Getting to know what triggers shame for you. There are certain aspects of our lives and things we do that we perceive to be a part of who we are. That may be parenting or physical appearance or creativity. Anything that you link to your self-worth can give rise to shame. In order to build and maintain a sense of self-worth, we need to understand that our worthiness as a human being is not dependent on living mistake-free.

- Reality-checking the criticism and all the judgements that follow. Whether it comes from someone else or inside your own head, judgements and opinions are not facts. They are narratives and stories that can significantly change our experience of the world. So looking after your self-worth involves taking the name-calling and personal attacks out of the equation and focusing on the specific concrete behaviours and their consequences. Reminding yourself that being imperfect, making mistakes or failing is all a part of being human. Making friends with your fallibility means that when you do fail you don't have to feel worthless. You can use each experience to your advantage by learning from it.

- Minding what you say. Criticism will always hurt a little; that is our brain doing its best to keep us safe. There is no antidote that makes it all OK forever and ever, criticisms bouncing off your armour. What use is armour anyway, when the worst critic is in your own head? A harsh comment or criticism can leave you feeling winded. Of course, naturally, you then spend the next five hours rehearsing it again in your head. The brain wants to pay attention to it because it is a threat. But each time you go over it in your mind, the stress response is triggered again. So one kick in the stomach can feel more like a hundred kicks in the stomach. Time spent thinking over a helpful criticism that we can put to good use and that adds to the work we are doing in the world is time well spent. Ruminating and churning over a nasty comment with no sense of how this can help you is just a continuation of the attack on your character.

- Talking to yourself in the right way after a criticism. This is vital if you want to have the capacity to move through shame and bounce back. When we are in shame we may feel a sense of self-loathing, and we convince ourselves that we need to continue the attack. The idea of responding to ourselves with respect and compassion feels undeserved and indulgent, as if it would let you off the hook and stop you bothering to try harder. But in reality, if you want a man to get up off the ground, you have to stop beating him. The key to using all criticism to your advantage is having your own back, having so much self-compassion that you are able to listen to criticism and decide which of that criticism you will take on board and use to your advantage as a learning experience, and which voices offer you nothing but dents in your self-esteem and crushed confidence.

- Talking about shame itself. Reach out to someone you trust and confide in them. Secrets, silence and judgement intensify shame. Sharing our experience with someone who responds with empathy helps us to leave shame behind and move on.

Understanding you

Living the life you want to live in the face of criticism means getting clear on:

- The opinions that truly matter to you and why. Whose opinions matter most to you? Saying 'I don't care what anyone thinks' is rarely true and hides a world of insecurities. It stops us creating meaningful connections with others because it closes off any avenue of communication in which both voices matter. But the list of whose opinions truly matter needs to be small. It is also worth pointing out that acknowledging who matters does not mean it is your responsibility to please them. It just means you are willing to listen to their feedback, even when it is not praise, because you know it is likely to be honest and in your best interests, therefore most likely to be helpful.

- Why you do what you do. The one person you most need the approval of is you. When the way we are living is out of line with our values and what matters most, life stops feeling meaningful or satisfying. Understanding the kind of person you want to be and how you want to live your life, how you want to contribute to the world, is the road you want to stay close to. When you know exactly who you are and who you want to be, it is much easier to choose which criticisms to take on board and which ones to let go.

- Where those familiar critical voices are really coming from and whether they are warranted and helpful or detrimental to our wellbeing. When there is someone in your life who is predictably critical, you hear their voice before they even say anything. Over time we internalize their constant criticism so that it becomes the way we speak to ourselves. So we may be highly self-critical because we have learned to be. Recognizing that we learned that way of speaking to ourselves helps us to acknowledge that we can re-learn a new internal dialogue that serves us better.

160

Chapter summary

- Learning the skills to deal with criticism and disapproval in a healthy way is a crucial life skill.
- We are built to care what others think of us, so telling ourselves that we don't care is not the answer.
- People-pleasing is more than just being nice to people. It is the persistent placing of others' wants or needs above your own, even to the detriment of your own health and wellbeing.
- Understanding why some people are highly critical helps.
- Nurturing your own self-worth and resilience to shame is both possible and potentially life-altering.

CHAPTER 19

The key to building confidence

As a teenager growing up in a small town, I thought of myself as a confident person. Then I left that small town and went off to university over 100 miles away. But much of my confidence, which I thought was part of me, stayed at home. I was vulnerable, unsure of myself, not clear on who I needed to be to fit in. Over time, university life became the new normal and I built up my confidence, brick by brick, once again.

After I graduated I got a job as a researcher for an addictions service. Feeling confident to meet the demands of university was no longer enough. I needed to tolerate that vulnerable feeling once again in order to build up confidence in this new arena. The same happened when I began my clinical training, then after qualification, after the birth of my first child, when I started my own practice and again when I started making my work public on social media.

With every turn I have ever made, the confidence that once seemed enough suddenly seems inadequate and vulnerability returns. Confidence is like a home that you build for yourself. When you go somewhere new, you must build a new one. But when we do, we're not starting from scratch. Every time we step into the unknown and try something new, experience that vulnerability, make mistakes, get through them and build some confidence, we move on to the next chapter with evidence that we can get through tough challenges. We bring with us the courage we need to take that leap of faith again and again. That same leap of faith the trapeze artist must take every time she lets go of one bar before grabbing the next one. She is always vulnerable, never completely safe, but each time she tries, she knows she can meet that risk with the courage needed to make it happen.

To build confidence, go where you have none

Confident is not the same as comfortable. One of the biggest misconceptions about becoming self-confident is that it means living fearlessly. The key to building confidence is quite the opposite. It means we are willing to let fear be present as we do the things that matter to us.

When we establish some self-confidence in something, it feels good. We want to stay there and hold on to it. But if we only go where we feel confident, then confidence never expands beyond that. If we only do the things we know we can do well, fear of the new and unknown tends to grow. Building confidence inevitably demands that we make friends with vulnerability because it is the only way to be without confidence for a while.

But the only way confidence can grow is when we are willing to be without it. When we can step into fear and sit with the unknown, it is the courage of doing so that builds confidence from the ground up. Courage comes first, confidence comes second. This doesn't mean that we have to dive in at the emotional deep end and risk overwhelming ourselves.

But it does mean that we must recognize how fear helps us to perform at our best and that we need to change our relationship with that fear so that we no longer need to eliminate it before we try. We learn to take fear with us.

Overleaf is the Learning Model (Luckner & Nadler, 1991) that we can use as a guide for building confidence. Note down what aspects of your life might be in the comfort zone, what tasks feel challenging but manageable and which things you would put in the panic zone. Every time you step into the stretch zone, you are doing the work of building your confidence by flexing your courage.

When you are trying to build self-confidence, it is a process of building self-acceptance, self-compassion and learning the value in vulnerability and fear. It is often a balancing act that doesn't always feel easy. Along the way, all of the tools in this book can be used, as they each help to build up your capacity to both lean into effort and tolerate the discomfort, then pull back and replenish.

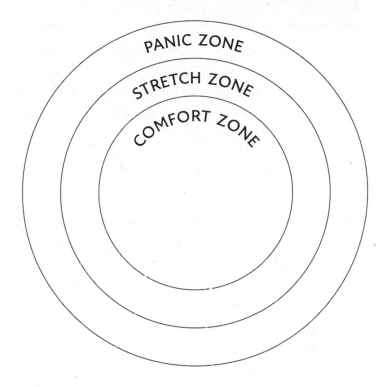

Figure 8: The Learning Model (Luckner & Nadler, 1991).

Some of the main ingredients for that leap of faith we need for stepping into the stretch zone are:

- Recognition that you can improve with effort.
- Willingness to tolerate the discomfort of being vulnerable for a while.
- The commitment to yourself that you will always have your own back and do the best by yourself whether you

succeed or fail. This means embracing self-compassion as a life practice and being your own coach, not your worst critic.

- Understanding how to move through the shame that can arise from failure to prevent our tendency to quit on our dreams in an effort to avoid the shame of a setback. See Section 3 for more on this.
- To build confidence, we don't have to live in fear. We have to develop a daily pattern of stepping into fear, sitting with it, and stepping back out, giving ourselves time to recover and replenish ready for tomorrow. See Section 6 on fear.

Why you don't need to work on your self-esteem

There is a whole industry built on the concept of self-esteem and the idea that if we can just believe in ourselves, we will perform better and improve our relationships and our overall happiness in life.

Self-esteem generally means being able to evaluate yourself positively and believe in those appraisals (Harris, 2010). So anyone trying to help you increase your self-esteem might ask you to list what you like about yourself and what your strengths are, and try to convince you to believe that you can become a 'success' in the world. But we have a problem with the concept of 'success'. We link it to the idea of wealth, winning, standing out and being acknowledged by others. So how do you know if you're winning? You compare yourself to others. Maybe you go online and pick any of the 4.6 billion internet users from across the world. With a pool that size, you are guaranteed to find someone who is doing better than you at something. When you do, self-esteem might take a hit because if you are not the winner, you can start to see yourself as a loser instead.

What if you stay offline and only compare yourself to your friends and family instead? Doing that is not going to nurture healthy relationships. Associating a measure of 'success' to worthiness would inevitably make it difficult to truly connect with the people you are comparing yourself to. What happens when you lose your job and your friend gets a promotion? A review of the research by a group of psychologists showed that high self-esteem is not linked with better relationships or better performance. But it does correlate with arrogance, prejudice and discrimination (Baumeister et al., 2003). They found no significant evidence that trying to boost self-esteem through intervention had any benefits.

Self-esteem cannot be relied on when it is contingent on being a 'success'. It is psychological rent that you can never stop paying. The second you notice signs that you might be less than, you brand yourself as not enough. So you keep running on a hamster wheel of success, driven by a scarcity mindset and the fear of being inadequate.

Ditch the positive affirmations

You cannot open a social media platform without coming across an offering of daily affirmations. The idea behind this is that if you say something to yourself enough times you will start to believe it and become it. But as it turns out, things aren't quite that simple. For those who already have high self-esteem and believe in themselves, repeating affirmations can have a small benefit of feeling a little better. But some studies have shown that for people with low self-esteem, repeating affirmations and statements that they don't believe, for example, 'I am strong, I am lovable,' or being asked to focus on all the reasons that statement is true, tended to make them feel worse (Wood et al., 2009).

The reason for this may be that internal dialogue we all have. If you say out loud that you are strong and lovable, but you don't believe that you are, then your inner critic will get to work coming up with all the reasons why you are not strong or lovable. The result is an internal battle, and plenty of time to focus on all the narratives that bring you down while you are desperately trying to push them away.

So what works instead? Well, the study I previously mentioned found that when those with low self-esteem were told it is OK to experience negative thoughts, their mood improved. They no longer had to battle with trying to convince themselves of something they didn't yet believe. Therefore, on the days when we don't feel strong, we don't need to tell ourselves that we are. We can acknowledge that feeling this way some-times is a part of being human. We can respond to it with compassion and encouragement. We can then turn to the things that help us to feel confident in our own strength again, by using all the tools at our disposal to move through tough times in line with the person we want to be. The way that we start to believe something more positive about ourselves is to use action to create evidence for it.

While affirmations may not be the best strategy for those with low self-esteem, words still matter. If mistakes and fail-ure lead to an onslaught of self-attack, do not let any of it go unchecked. Professional athletes have professional coaches for a reason. In day-to-day life we don't have that, so we must be that voice for ourselves. The natural emotional response to failure influences our thoughts and makes us more vulner-able to self-criticism. So we can't always stop it, but we can respond to it with an alternative that serves us better. To build confidence, you have to get to work being your own coach instead of your own worst critic. That includes respond-ing to failure in a way that will help you to get up, dust yourself off and get back out there. A professional coach would not bully you with words, or chant affirmations that you couldn't believe in. They bring honesty, accountability, unconditional encouragement and support. They are in your corner, what-ever the score, with your best interests at heart. Doing that for yourself is not always easy, but is a life skill that we can improve on with practice.

Toolkit: Changing your relationship with fear to build confidence

To begin building confidence in something that makes you feel nervous, you can practise welcoming the sensations of fear and sitting with them, without pushing them away. To do this, we don't have to put ourselves in a situation that causes intense panic or terror. In fact, that is not advisable. Instead, we can practise by dipping a toe in the water. Stepping slightly out of that comfort zone just enough that you can feel the stress response without feeling overwhelmed.

- Write down the situation in which you would like to build more confidence. Put the most vulnerable situation at the top. Then list any variations of that situation that may feel more manageable but still challenging. For example, if I want to build my confidence in social situations, I may put parties at the top of the list as the occasion in which I feel least confident. Slightly easier than that might be a party at which I know everyone. Easier than that might be a small gathering of close friends. Easier still might be going to a café with a trusted friend. Once you have your list, you don't start at the top. You take the scenario that presents a challenge but still feels possible. You then repeat that behaviour as often as you can. Once your confidence grows and that situation becomes a comfort zone, you move up to the next thing on the list.

'The perfect nurturer' is a tool originally developed by Paul Gilbert and Deborah Lee and used in Compassion Focused Therapy (CFT). It can be a helpful way to turn your focus towards the self-talk you need when building your confidence.

- A perfect nurturer is an image of a person that you can return to in order to feel safe and nurtured when that is what you need. If you prefer the idea of a coach, you can use that instead.
    - Create an image in your mind of the perfect nurturer or coach (this could be a real or imagined person).
    - Imagine you are sharing with them what you are currently facing, how you feel about it and what you want to work on.
    - Take some time to imagine in detail how that perfect nurturer or coach might respond, and write it down. This sets the tone for the words you can start using to respond to yourself as you work on building your own confidence and inevitably face vulnerability along the way.

Chapter summary

- Confidence cannot grow if we are never willing to be without it.
- To build confidence, go where you have none. Repeat every day and watch your confidence develop.
- Confidence is situation-specific, but what you keep as you move around is the belief that you can tolerate the fear as confidence grows.
- You do not need to overwhelm yourself with your worst-case scenario. Start with small changes.
- Along the way, be your own coach, not your worst critic.
- Courage comes before confidence.

CHAPTER 20

You are not your mistakes

Most self-doubt is linked to the relationship we have with failure. I am not about to sit here and tell you to just be OK with failure, then everything will be easy. That is not true. Failure is never easy. It stings every time. We all want to be enough. We all want to be acceptable and failure is a sign that maybe we weren't enough this time.

It is not only our relationship with our own failure that needs to change, but also how we respond to the failure of others. You don't have to spend too long on Twitter to develop an overwhelming fear of failure. Say the wrong thing in a tweet and a collective army of tweeters invade and hound you with verbal abuse, demanding you be pulled down from whatever heights you may have climbed in your lifetime. I have seen this happen to people who make simple mistakes in their use of language and immediately apologize for it. Given that social media is a magnified reflection of who we are as a society, this speaks to me about the intense shame we associate with any form of failure. Those who are highly self-critical are more likely to be critical of others. If we believe that mistakes and shortcomings should be met with humiliation and shame no matter what the intention, how do we ever begin to be OK with taking risks and making mistakes ourselves?

Something that has helped me a lot is understanding that how other people respond to my failures does not provide an accurate assessment of my personality and worthiness as a human, but instead indicates how that person relates to failure. Accepting failure is hard in environments where people attack each other for mistakes. No matter how hostile we are towards failure collectively, changing our relationship with failure must begin with us. Failure hurts every time, whether the environment is safe or not. Therefore, we avoid it at all costs. We quit when things get hard, we switch to an easier, safer option, or we refuse to begin at all. All of those choices are addictive because they bring that blissful sense of relief. Phew! I don't have to face that today. We do that enough times and it becomes a pattern in our lives that keeps us stuck in a comfort zone that feels like treacle, with all its lethargy and lack of energy for anything.

If the opposite of resisting failure is accepting it as a part of growth and learning, how do we do that? It is one thing to communicate something intellectually, it is quite another to feel it and truly believe it in the moment. Saying it is only useful if you can buy into it. Belief is everything. So we need to say something that we can buy into. Therefore, it is no use trying to convince ourselves that it is safe to fail. We cannot guarantee

how others will react. There will always be critics. Not everyone is going to help you up when you fall. Our only option is to
commit wholeheartedly to doing that for ourselves. Start by
recognizing that coming back from failure cannot be dependent on others. Using support that is available is always a good
idea, but we cannot always count on someone else being there,
so committing to ourselves that we will take responsibility to
tend to our wounds with compassion and dust ourselves off
after a fall is essential if our resilience is not to be dependent on
others doing that for us.

Bouncing back from failure

1. Recognize the sensations in your body and signs in your
   urges and actions that indicate to you how you are
   feeling. If you find yourself using all of your favourite
   numbing activities – hours of TV, alcohol or social
   media. The sting of failure drives us to block. So even if
   you don't initially notice the feeling, you can notice the
   blocking behaviour.
2. Get unstuck. Remember the story about Jim Carrey
   removing the mask? It had less power over him when it
   was not right in front of his face. We can do this with
   emotion by seeing it as an experience washing over us,
   not as who we are. If you can label the emotion, it
   enables your mind to take a step back from it. If you
   can label the pattern of your thoughts, that will do the

same thing. The account of what is going on, as told by your mind, is not a fact, but a theory, opinion, story or idea. That opinion is tinted with critical voices from your past and present, memories of feeling vulnerable or having failed. As we get to know the patterns of those critical voices and where they may have come from, we can even give that stream of thought a name. There is Helga, giving her two cents! Doing this can be powerful at helping us to get some distance from the self-attack and giving us more choice over whether we buy into it as factual or see it as just one (very unhelpful) option.

3. Notice the urges to block the painful feelings and keep reminding yourself that you do not have to act on those urges. When we stop fighting emotion, but instead allow it to wash over us in all its might, it is painful and it is messy. But it also washes past. If we try to push it down and hold it under, it stays there, waiting for its chance to be processed. The opposite of blocking emotion is to be curious about it. Move towards it. Observe and notice the whole experience, all while doing Step 4.

4. Soothe your way through and have your own back with the commitment of the best friend you could ever wish for. Be honest with yourself while also offering unconditional love and support. 'Wow, that was tough. Hold on.' The best kind of friends know they cannot fix things for us. But they stand strong by our side through the whole thing.

5. Get learning. Anyone coaching a professional athlete will analyse each performance. They don't only look for what went wrong. They also look at what is working. So when the pain of the setback has been calmed, get to work on making that experience useful for you. Don't neglect to notice the things you did well. Appreciate what worked and what didn't. Be your own coach so that you can learn from the experience and move forward.

6. Return to what matters. Failure and setbacks hurt every time but dusting off and getting back out there may still be in line with our values. When the pain of failure is still present it can be difficult to even consider trying again. Instead we want to run and hide. Coming back to your values and the reason you are doing this thing will help you make a decision based on your best interests and the life you want to have, rather than basing that decision purely on the pain. That being said, I don't want to trivialize just how much we can be overwhelmed with emotion after failure, so take your time. It is important to work through the experience first, and try again when you are ready.

Check out the chapter on values for more details (see page 284), but in the heat of the moment we don't always have time to whip out our worksheets and double-check what is in line with our values. In that moment, simply ask yourself, 'When I look back on this time, what choices would I be proud of? What actions could I take that I will be grateful I took a year from now? How can I learn from this and keep moving forward?'

Chapter summary

- Most self-doubt is linked to the relationship we have with failure.
- How other people respond to your failure does not say anything about your own personality or worthiness.
- The sting of failure can drive us to numb or block emotion. So even if you don't initially notice the feeling, you can notice the blocking behaviours.
- Be your own coach and turn failure into a learning experience as you keep moving forwards in line with what matters most to you.
- The emotional response to failure can be overwhelming, so take your time.

CHAPTER 21

Being enough

The brick wall that most people hit on their way to self-acceptance is the misconception that it will cause laziness and complacency. They think that self-acceptance means believing you are OK just as you are, therefore you will have no motivation for improvement, work, achievement or change. In reality, the research shows us that those who develop self-acceptance and learn to be self-compassionate are less likely to fear failure, more likely to persevere and try again when they do fail and generally have more self-confidence (Neff et al., 2005).

Self-acceptance and the compassion we show ourselves when we are accepting of the self is not the same as becoming indifferent to the world and passively resigning yourself to accept defeat when things are tough. Having unconditional love for yourself means doing the opposite sometimes. It can mean taking the more difficult road because it is in your best interests. It is refusing to kick yourself while you are down or indulging in self-loathing, and instead using every ounce of strength to pull yourself back up after a fall.

The difference is that when you strive, you do so from a place of love and contentment rather than striving from a place of fear and scarcity.

If we don't do the work to develop self-acceptance, we set ourselves up to live a life in which we may need constant reassurance, get trapped in jobs we hate or relationships that cause us harm, or find ourselves living with resentment.

So how do we start developing self-acceptance?

Understanding yourself

Sounds simple, but many people go through life without much in the way of examining their own patterns of behaviour that impact on their experience of life. To build self-acceptance we first need to understand who we are and who we want to be. That comes from practising self-awareness. We become self-aware through self-reflection. Keeping journals, going to a therapist or talking to friends can all help us to reflect on ourselves and our experience in a way that enables us to learn more about who we are and why we do what we do. Self-acceptance involves listening to our own needs and meeting them. If we are not paying attention, we don't always pick up on the signs.

During that process it is important to be attentive to the parts of us that we feel proud of and the parts of us that we may prefer not to think about – the things we dislike, feel anxious or regretful about, or want to change. But when we do reflect on those more difficult aspects of the self, it is crucial that we come at it with the compassion of an observer if we are to learn. If reflecting on difficult situations triggers intense emotions that make it harder to think clearly, it can be helpful to get support from a therapist to help us work through it.

Painting the picture of self-acceptance

Let's say that from the moment you closed this book you started living your life with unconditional self-acceptance. What would that look like? What would you do differently? What would you say yes to? What would you say no to? What would you work harder on? What would you let go of? How would you speak to yourself? How would you speak to others?

Try writing down your answers to those questions in great detail and create a vision of how the idea of self-acceptance would translate into behaviour change for you. As with most changes, the action comes first and the feeling comes later. So living a life in which you feel a sense of self-worth means making it a life practice. The work is never done. You never arrive. You do the work every day to live in line with unconditional self-acceptance.

Accepting all of you

While you maintain a sense of self that stays with you throughout your life, you also experience a wide range of emotional states that are constantly in motion, changing from one moment to the next. We take on different roles and behaviours in different scenarios and many people see those as different parts of themselves. Depending on our early life experiences and how the world responded to those emotion states, we can feel that some parts of the self are less acceptable than others. If anger was unacceptable for you growing up, then it might feel much harder to treat yourself with compassion and acceptance when you are feeling angry. This makes self-acceptance conditional on how you feel.

 Try this: You can build your awareness of how you respond to various emotions and practise stepping back from them and responding to them with compassion using the following exercise that is used in Compassion Focused Therapy (CFT) (Irons & Beaumont, 2017).

Spend a few moments thinking back to a recent event that triggered a mix of emotions. It is a good idea to start with something that is not too distressing so that you don't become overwhelmed while trying to practise the exercise.

1. Write down a few thoughts about the event.
2. Write down the different emotions it brought up, for example anger, sadness, anxiety.

3. For each emotion that you identified, take one at a time, connect with that feeling and explore your answers to the following questions.

   1. Where do you notice that feeling in your body? How did you know that feeling was there?
   2. Which of your thoughts were linked to that feeling? If that emotion could speak, what would it say?
   3. What urges came along with that feeling? If that emotion had been able to decide the outcome, what would it have made you do? (For example, anxiety may have wanted you to run away, anger may have wanted you to shout at someone).
   4. What does that part of you need? What would help that emotion to come back down to calm?

Once you have done this for each emotion that was present, end by answering the questions for your compassionate self. The part of you that wants to show yourself unconditional love and acceptance.

As you do this for each emotion, give yourself time to step back from connecting with an emotion before stepping into the next one (if there were lots of mixed emotions). Each time you do this, you are strengthening the ability to defuse those emotions and get a good understanding of them without getting overwhelmed by them.

This can be a helpful exercise for examining those mixed emotions, allowing us to see that even those emotions that we once saw as unacceptable are just normal. Each reflects a different way of interpreting the situation, so we can come to different conclusions about which direction to take next. Taking time to get a birds-eye-view of emotional experiences like this can help us to engage with our compassion even in situations where we have been taught to be harsh on ourselves.

Kicking self-criticism

- What does your self-criticism sound like? What words do you use?
- What does it focus on?
- What types of things do you criticize yourself for? Appearance, performance, personality traits, comparison?
- Certain forms of self-criticism can be more damaging than others.
- Sometimes self-criticism can take the form of telling yourself that you are inadequate after a failure.
- But when it goes further than that and you feel a sense of disgust and hate towards yourself, it is even more pervasive and shaming.

 Try this: This is a quick exercise that can help us get some distance from our inner critic and see it for what it is.

Once you have reflected on all the different ways you criticize yourself, take a moment to imagine that critic as a person outside your own head. What would they look like? What would their facial expression and tone of voice be as they spoke to you? What emotions would they be expressing? How does it feel to have them in front of you? What do you think might be their intention? Is this a misguided attempt to protect you in some way? Is this someone you would want to spend time with? Is this someone who could help you to live a happy life? And finally ask yourself, what is the impact of spending every hour of every day with that critic?

Finding your compassionate side

When your inner critic has been a close (but unwanted) companion for most of your life, it is almost impossible to simply decide to remove them. That well-practised action means those neural pathways are easily accessible for your brain. So that voice will speak up from time to time. What we need to do is provide ourselves with a new, healthier and more helpful voice, and then start practising. In the same way that you made time to see and hear the inner critic, let's invite the compassionate side of you to the party. The side that wants the best for you and recognizes the harm that the self-attack causes. That part of you still wants you to grow and achieve, but from a place of love rather than shame.

Take some time to consider what compassionate self-talk might sound like. Remember, this is not the same as positive thinking. Someone who has compassion is honest and kind, encouraging and supportive, and wants the best for you. What words would you use when showing compassion to others? What words of compassion have others expressed to you? Bring to mind a memory of a time when someone showed you compassion. How did they look at you? What did they say? How did that make you feel? What would it be like to have access to that voice at any time?

 Try this: To strengthen our compassionate side, we simply have to put in some reps on a regular basis and practise. Try writing a compassionate letter to yourself. Allow yourself to write spontaneously in the way you might write to a close friend who was suffering or trying their best to change. How would you express to them that you had their back always and wished for their suffering to ease? Nobody needs to read the letter. But the process of engaging with your compassionate self and thinking through the various ways you can express that helps you to build that mental muscle to use it when you most need it.

If you find yourself struggling to access that feeling of compassion for yourself, focus on someone that you love unconditionally and imagine you are writing to them, or use the words that loved ones have said to you in the past.

Chapter summary

- There is a misconception that self-acceptance will cause laziness, complacency and lack of drive.
- In reality the research shows us that those who develop self-acceptance and learn to be self-compassionate are less likely to fear failure and more likely to try again when they do fail.
- Self-acceptance is not the same as passively accepting defeat.
- Self-compassion often involves taking the more difficult road that is in your best interests.

6

On Fear

CHAPTER 22

Make anxiety disappear!

For as long as I can remember, I have had a fear of heights. For the most part I was able to avoid them, growing up. But when I met my now husband, we went on a trip to Italy together. We went to visit the Leaning Tower of Pisa and as we stood looking at it, he held out two tickets. We were going to the top. A deep intake of breath was followed by another glance up at the tower that leans at 3.99 degrees and looks worryingly close to falling over.

My heart started to beat against my ribcage and I felt sick. But the tickets were bought, so up we went. To get to the top of the Leaning Tower of Pisa, you climb the narrow stone steps that spiral around inside the tower. The ground beneath your feet is not level and as you scale the tower you get the sensation that it is starting to fall over. At least that was my assessment at the time. With a queue of people behind me I kept going. Once at the top the tilt feels even worse. Everyone walked straight over to the edge to take in the view. I had the overwhelming urge to get closer to the ground. I moved as far away from the edge as I could get and dropped to the ground. There was a vague attempt to make it look like I was just sitting down casually to have a rest, but by now my fear of embarrassment was outweighed by my fear of falling to my death. Of course, sitting on the floor made me no safer. But it was not based on logic. My brain was sending out strong signals to the rest of my body to get down. I even found myself looking down at the slab of stone to avoid looking out at the view. Matthew took a picture of me crouching on the floor and we now have a funny memory to look back on. But what was going on at the time? Why did I have the urge to get on the floor?

My phobia, learned early in childhood, meant that the moment I saw those tickets and imagined climbing to the top of the tower, my body reacted. My heart started beating faster, my breathing became shallow and fast, my palms sweaty. The building being off balance only contributed to my predictions that I would fall to my death at any moment. But the brain's alarm system, which was sounding and telling me to get to safety, is more like a smoke alarm. It doesn't have time to take account of all the facts. Its job is to sense danger and let me know. It takes its information from the distress signals in my body, information about my surroundings from my senses, and pieces that together with memories of what was happening when I last felt this way. In the same way that a smoke alarm can go off when there is a fire, it can also go off when someone is making the toast. The strong urge to drop to the ground was a suggestion from my brain that my body took very seriously (much to the amusement of everyone else on the tower). The fear was overwhelming, so I did the first thing I could think of to feel safer. I wanted the fear to go away.

Those strong urges to get to safety are not a fault, but the brain working at its best to keep us safe. Crucially, my actions did not make me any safer but just helped me to *feel* safer.

The question of how to make anxiety go away might be one of the most common questions I receive. It makes sense to ask that. Anxiety is uncomfortable at best, overwhelming at worst. When you feel anxious, your body is working really hard so it's also exhausting. Nobody wants to live with anxiety every day.

Here is where I went wrong on the Leaning Tower of Pisa and did very little to help my fear of heights. I avoided the fear as much as I could. I got close to the ground, I avoided looking out at the view. I even shut my eyes when I could. I tried to convince myself that I was not up high. I left the tower before my fear had left me. As my feet hit the comforting grass of the Field of Dreams once again, I felt a rush of relief and my body immediately calmed. My brain said, 'Phew! That was dangerous. Let's never do *that* again!' I did everything I could to make the fear go away as soon as possible. But all the things that give us that instant relief tend to keep us stuck in the long term.

If I knew then what I know now, and I was on that trip to tackle my fear of heights, here's what I would have done. I would have gone to the top of the tower and looked out at the view. The feelings would all have been the same, but this time I would allow it to be present without trying to avoid it. I would have responded to them by taking control of my breathing and I would have focused on breathing slowly. I would have reminded myself that my body and brain were responding in this way because I have memories of being up high and feeling unsafe as a child. I would have reminded myself repeatedly that I was, in fact, safe. I would turn my attention to why I was there and I would continue to breathe slowly for however long it took for my body to exhaust itself. When my fear started to subside and my body calmed, only then would I step down again. I would then repeat that pattern for as many days as possible, knowing that over time my body would habituate to the situation and my fear response would gradually reduce in its intensity.

Fear is a part of our survival response. It is supposed to be intensely uncomfortable, and the urges to escape and then avoid the feared situation are supposed to be strong. If we are in a survival situation, that system works incredibly well to keep us safe. That moment you are crossing the road and you hear a car horn too close, before you can even think about the situation, you dash to the curb quicker than you ever thought you could move. Then you feel the surge of adrenaline wash over your entire body. That is your fear response working at its best. But a system that works that fast does not have time to contemplate what signs are valid and which ones might be less reliable signs of danger. It senses, it acts. You survive. 'Thank you, brain.'

In other situations when your life is not in danger, the urges are still the same. You're asked to speak in a meeting and your heart starts pounding. Your heart may be getting your body ready to be alert and perform. But if you interpret that as fear and make your excuses to leave the room, then avoid those meetings in the future, you never get to experience talking in meetings and having it go well.

The things that give us immediate relief from our fear tend to feed that fear in the long term. Every time we say no to something because of fear, we reconfirm our belief that it wasn't safe or that we couldn't handle it. Every time we cut something out of our lives because of fear, life shrinks a little. So our efforts to get rid of fear today mean that fear gets to take over our life choices in the long term.

Our attempts to control fear and eliminate it become the real problem that dictates our every move. Fear is around every turn, in every novel situation we face, in every creative endeavour and every learning experience. If we are unwilling to experience it, what are we left with?

Chapter summary

- It is understandable to want anxiety to disappear. It is supposed to be uncomfortable.
- To fight fear you must first be willing to face it.
- Escape and avoidance only provides short-term relief but feeds anxiety in the long term.
- Our attempts to control fear and eliminate it become the real problem that dictates our every move.
- A threat response needs to work fast, so it tends to sound the alarm before you have a chance to think things through in more detail.

CHAPTER 23

Things we do that make anxiety worse

When we feel anxious about something, the most natural human response is to avoid it. We know that if we stay away, we'll feel safe, for now. But avoidance not only maintains anxiety, it makes it worse over time.

Your brain learns like a scientist. Each time it has an experience, positive or negative, it clocks that as evidence for its beliefs. If you avoid the thing you fear, you never give yourself the chance to build up evidence in your mind that you can get through it and survive. Just telling your brain that something is safe is not enough. You must experience it.

Your brain will take some convincing, so you need to repeat that behaviour over and over. As many times as it takes. The things you do most of the time become your comfort zone. So, if you want to feel less anxious about something, do it as often as you can. Use the skills to help you sit with the anxiety and it will reduce over time.

When we learn to face the things that make us feel afraid, we get stronger. When we do that day after day, over time we develop a sense of growth. Imagine if, over the next five years, you made your decisions based on the life you want to have, instead of fear.

We avoid the discomfort of fear in so many ways. If you feel anxious about a social event, you may avoid it by not going. Or maybe you do go, but you drink to excess before you go. Drinking may dampen that anxiety in the moment, leaving you feeling the need to do the same at the next social event. These safety behaviours work in the same way to numb the anxiety in the moment. But they don't help us to feel less afraid in the future. In fact, they do the opposite. They feed the anxiety for the future and we become dependent on using those safety behaviours, making life even harder.

Here is a list of some common safety behaviours that ease anxiety in the moment, but keep us stuck in the long term:

Escape – Whether it be in a social situation, the supermarket or a confined space, when anxiety hits we have the urge to get out of there as quickly as possible.

Anxious avoidance – The moment you say no to that invitation to avoid the social situation or opt for food deliveries to avoid the anxious feeling you get in the supermarket, you are rewarded with instant relief. 'Phew. I don't have to face that feeling today.' But the longer you stay away from something, the more the fear seems to grow. Then the day comes that you need to face it once again and it now feels overwhelming.

Compensatory strategies – This can happen after experiencing a high anxiety state. For example, someone with a fear of contamination or sickness may wash excessively after being in a hospital setting.

Anticipation – Also called sensitization, this is when we rehearse and anticipate various worst-case scenarios that may occur in a feared situation. We are often convinced that it is helping because it will protect us if we are prepared, but it can lead to hyper-vigilance and excessive worry without constructive planning, which leads to increased anxiety.

Reassurance seeking – In moments of anxiety and doubt we may ask for reassurance from a loved one that everything will be OK. It is hard to see a loved one in distress, so they are often more than willing to use reassurance to help calm the anxiety. But over time that instant relief can become addictive and we develop a dependency on that other person. We may need almost constant reassurance, or feel unable to leave the house without being accompanied by the person who makes us feel safe, which can weigh heavily on a relationship.

On Fear

Safety behaviours – We can also come to rely on things that we associate with safety if we don't trust ourselves to be able to cope when anxiety hits. We may feel unable to go anywhere without 'just in case' medications, or we take a mobile phone everywhere because looking down at it enables us to avoid conversation at social events.

Chapter summary

- When we feel anxious about something, the most natural human response is to avoid it.
- But avoidance maintains anxiety.
- Just telling your brain that something is safe is not enough. You must experience it to truly believe it.
- Your brain will take some convincing, so you need to repeat that behaviour over and over.
- The things you do most often become your comfort zone.
- If you want to feel less anxious about something, do it as often as you can.

CHAPTER 24

How to calm anxiety right now

If you struggle with anxiety you are probably hoping for a tip that you can use right now. Something easy to learn that will have instant effects. Many people feel this way at the beginning of therapy. That is why I always teach people this first skill as early as possible. It is easy to learn and takes just a few minutes to bring down the intensity of anxiety. At the very least, it prevents anxiety from escalating into panic.

When anxiety is triggered, you start breathing more quickly. This is your body's way of getting in extra oxygen to fuel the survival response.

You feel as though you cannot catch your breath. So you breathe faster with rapid, shallow breaths, then you have an excess of oxygen in your system. If you slow your breathing down, you can calm the body and, in turn, slow your breathing. Not only this, but if you can extend the outbreath so that it is longer or more vigorous than the inbreath, this helps to slow your heart rate down. When the pounding heart comes down, so does the anxiety response.

Some people like to count the breaths when doing an extended outbreath, such as breathing in for a count of 7 and out for a count of 11, or a variation that works for you.

Taking some time to practise slow breathing techniques is a great investment of time. It's an anxiety management tool that works in the moment. You can do it anywhere, any time and nobody even needs to know you are doing it. One of my favourites is square breathing. Just follow the steps below.

 Toolkit: Square breathing

Step 1. Focus your gaze on something square: a nearby window, door, picture frame or computer screen.

Step 2. Focus your eyes on the bottom left-hand corner and as you breathe in, count to 4 and trace your eyes up to the top corner.

Step 3. Hold your breath for 4 seconds as you trace your eyes across the top to the other corner.

Step 4. As you breathe out, trace your eyes down to the bottom corner, counting to 4 once again.

Step 5. Hold for 4 seconds as you move back to the bottom left corner to start again.

So, you breathe in for 4 seconds, hold for 4 seconds, out for 4 seconds and hold for 4 seconds. Focusing on something square can act as a guide and help you to keep your attention on the breathing, minimizing the chances of being distracted too soon. If you try it for a few minutes and feel like it's not working yet, keep going. It takes some time for your body to respond.

One extra tip is to practise this every day, at times when you don't feel anxious. When something is well practised, it is much easier to use when you find yourself overwhelmed with fear.

Movement

Another tool that has almost instant effects, and requires very little practice to master, is exercise. When your anxiety response is triggered, your muscles fill up with oxygen and adrenaline, ready to move fast. If you don't move and burn off that fuel, your body is like a rocket with engines firing and nowhere to go. Bring on the trembling, shaking, sweating and the urge to pace around the house.

Exercise is one of the best anxiety management tools because it follows the natural course of your threat response. Your body is geared up to move. Allow it to do that and your body can use up the energy and stress hormones it has produced and rebalance.

If you have a stressful day, try adding in a short jog outside, or an intense half hour with a punch bag. Physical movement will truly relieve your body of the physical stress so that when you sit down to relax, you can feel calm and fall asleep more easily, helping you to replenish further.

My extra tip here is that exercise is also a powerful prevention tool, so try to exercise even on the days when you don't feel anxious. This way you are setting yourself up to have a better day tomorrow. Your mental health will thank you for it.

Chapter summary

- When anxious our breathing becomes faster and each breath is more shallow.
- To calm the body take slower, deeper breaths.
- Try to make the outbreath longer and more vigorous than the inbreath.
- Give it some time and the anxiety response will begin to drop.

CHAPTER 25

What to do with anxious thoughts

Like many other kids in the early 90s, I was allowed to stay up a little later on a Friday evening to watch *Casualty*, a television show about an Accident and Emergency department. This particular episode (the only episode I remember to this day) was about a man who lived on the sixth floor of a block of flats. A fire breaks out downstairs and he is trapped. Not long later I am lying in bed, running through it in my mind. *What would I do if my house caught on fire? Is the house on fire now? How would I know? What if I wouldn't wake up in time? Maybe I should try and stay awake. Maybe I should open the bedroom door and check downstairs.* I lie there, eyes wide open, trying out different scenarios in my head. Imagining waking up my younger sister, who was in the room with me, opening the door to a cloud of smoke, opening the windows and calling for help. Soon the warm glow through the glass panel above my bedroom door started to look more and more like the orange glow of fire. I remained still and silent, unable to move, listening for the crackling sounds and waiting for the smoke.

That night, I not only believed that a fire in my home was possible, I saw it happening over and over in my own head. I bought into every scenario as if it were occurring. Played it over and over in my head like a movie.

When a worrying thought pops into your head, it's like driving past a car accident: you can't not look at it. Thoughts of danger demand your attention for a reason. Your brain is offering up a story for what might be happening and if there is a chance of the worst-case scenario happening, then you had better be prepared.

As I explained in a previous chapter, your brain acts a bit like a smoke alarm. Every time you sense a threat in your environment, that alarm gets triggered and tells your body to get into survival mode. That is called your fight or flight response. Your body gears up to either fight off that threat, or escape quicker than you ever thought you could move.

A smoke alarm is made to go off when there is a fire. A necessary tool for survival. Just like the smoke alarm, anxiety can be triggered even when we are not truly at risk. But when you burn the toast and the smoke alarm sounds, you don't remove the smoke alarm. If you understand why it is there and how it works, you can start to work with it, make adjustments, open a window. You get the idea. We cannot remove our survival response. We would not want to. But we can learn about what exacerbates it and make adjustments so that we can detect a false alarm and act accordingly.

Get some distance

Thoughts are not facts. They are guesses, stories, memories, ideas and theories. They are a construct offered to you by your brain as one potential explanation for the sensations you are experiencing right now. We know they are not facts because they are so heavily influenced by your physical state (hormones, blood pressure, heart rate, digestion, hydration, to name just a few), by each of your senses, and by your memories of past experience.

So what does this mean for those anxious thoughts that pop into our heads? It means the power of that thought, and any other thought, is in how much we buy into it. How much we believe that thought to be a true reflection of reality. The best way to break down the power that thoughts have over our emotional state is to first get some distance from them. How do we get distance from something that is inside our head?

Getting distance from thoughts can be done in a number of ways. Mindfulness is a great skill to start practising for building that ability to be able to notice your thoughts and let them pass without getting caught up in them. But being aware of the types of thought biases that tend to arise when we are anxious also helps. If you can notice a thought for what it is – a biased guess – and label it as such, then doing that is one way of holding the thought at arm's length. Your mind can then see it as just one possible perspective. We are then in a much better position to be able to consider alternatives.

One way to get that distance from anxious thoughts is to use distanced language. This helps to turn the dial down on the emotion. Rather than saying 'I am going to make a fool of myself during this speech,' say, 'I'm having thoughts about making a fool of myself. I notice those thoughts trigger feelings of anxiety.' I know thinking or speaking in this way may feel awkward at first. But it makes a difference in helping you to step back from the thoughts and see them as an experience, not as you.

Another way to get distance from those thoughts, and my personal favourite, is to write them down. This is not exclusive to anxious thoughts. Any time that you want to get some distance and a new perspective on your emotional state or situation, write down everything you are thinking and feeling. Seeing what you have written on the page can be a powerful way to process and make sense of your experience from a bird's-eye view.

Spot biased thoughts that make you feel worse

There are a few thought biases that commonly occur when we are feeling anxious:

Catastrophizing

Catastrophizing is when your mind jumps to the worst possible scenario and offers it to you as a prediction of what might happen now. It plays that out for you like your own personal horror movie on repeat in your mind. It is one possible prediction, but not the only one. When we play it over and over in our mind and buy into it as an absolute certainty, anxiety goes up. In a previous chapter I mentioned my fear of heights growing up and my early attempts to face it. At the top of the Leaning Tower of Pisa, my thoughts that I was about to die were catastrophizing thoughts that were repeated over and over. It turns out they were just one possible ending to that story, as the actual ending went like this: I walked back down the steps and continued with my holiday.

Personalizing

Personalizing is when we have some limited or ambiguous information about the world and make it about us. For example, I'm walking down the street and see a friend on the other side of the road. I call her name and wave, but she doesn't wave back. Immediately, my personalizing thoughts tell me that she must hate me. I must have said something to offend her. Maybe all our friends have been talking about me and I thought I had friends but now I have none.

There are thousands of potential alternatives to the story my mind has offered as an explanation for this. Maybe she didn't hear me. Maybe she normally wears contact lenses. Maybe she has just had a huge fight at home and can't bear to speak to anyone in case she bursts into tears in the street. Maybe she was daydreaming. The list goes on and on. The personalizing bias demands our attention because it is threat-focused. If my friends suddenly hate me, this is something I need to focus on.

Mental filter

The mental filter is that tendency for us to hold on to all the information that makes us feel worse, and neglect all the information that could help us feel differently. Let's say you post something on social media and you get fifty comments. Forty-nine of those comments are positive and encouraging. One is negative and points out something that you already felt inse-cure about. The mental filter is when we focus our attention on that one negative comment and neglect to consider the other forty-nine. The mental filter was definitely at play when I focused in on the fact that the Leaning Tower of Pisa was lean-ing and failed to consider that it has remained standing for hundreds of years and has a large team of professionals con-stantly monitoring its safety.

The brain naturally wants to focus on threatening information because its job is to keep us safe. If we are already stressed or anxious, then the brain will do that even more. The brain receives the information from the body that all is not well, and starts scanning the environment (and your memory) for possible rea-sons. This is when the mental filter kicks into action. Your brain is on a mission to make sense of the anxiety symptoms. But, if we can notice the mental filter in action, we can recognize the bias in the information we are focusing on and intentionally choose to consider the other information available.

Overgeneralizing

Overgeneralizing is when we take one experience and apply it to all experiences. If you interview for a job and get turned down, overgeneralizing thoughts would sound like, 'I'm never going to get a job, so what's the point of applying for anything else?' or after a breakup, 'I screw up every relationship so I'm not going to date ever again.' Overgeneralizing makes anxiety worse for a couple of reasons. It leads to a more intense spike of emotion because it turns one problem into a bigger, life problem. Secondly, it often leads us to avoid the situation in the future, which feeds anxiety and makes it much harder to face.

Labelling

Labelling is similar to overgeneralizing but involves taking one event or period of time and using that to make global judgements about who you are as a person.

If you experience a period of anxiety in your life, and from that point label yourself as an anxious person, you start to form a concept for your self and your identity which then impacts on how you expect to feel and behave in the future. Each emotion, behaviour and period in our lives is temporary and not necessarily a reflection of who we are permanently.

So, when you notice that you are labelling yourself as a certain type of person, don't let it go unchecked. Doing so has an impact on the emotions that are constructed by your brain in the future. Instead, acknowledging the specifics of the experience as a temporary one helps to distance you as a person from the experiences that you encounter along the way. It's much harder to change an identity as an anxious person than it is to simply reduce anxiety.

Fact check

Since the power of any thought is in how much we believe it to be a true reflection of reality, 'thought challenging' can be a helpful process for many. If a thought is causing you distress, it makes sense to work out whether it is fake news or worth feeling so anxious about. Thought challenging is a simple process. When you start, it's easier to do it in hindsight, after the event has passed. If you notice anxious thoughts you can follow these steps to challenge them.

1. Write down the anxious thought.
2. Draw a line down the centre of your page to make two columns. Like a lawyer weighing up the facts, write down a list of all the evidence that the thought is a true fact. Evidence only counts if it would stand up in court as evidence.
3. In the second column, list all the evidence that the thought is not a fact.

4. If the exercise reveals the anxious thought to be less
   factual than you initially believed, then it is a good time
   to start considering alternative ways to think about the
   situation.

This exercise is very simple but can be helpful to just loosen our initial belief in the thought and open up the opportunity to consider alternative interpretations.

However, if you find this just leads to an internal argument about how true the thought is, that is when this technique becomes less helpful. If that happens, leave the thought-challenging exercise and instead use the other techniques that focus on distancing yourself from the thought.

Spotlight of attention

It's New Year's Day 2010. I am pulling on a crusty blue boiler suit and zipping it up at the front with my eyes closed. I take a deep breath, as if it's going to be the last breath I take. I feel sick. I wipe the sweat from my palms down the front of the boiler suit. I open my eyes and Matthew is grinning at me.

'Ready?' His smile is so wide he looks like he has a hanger in his mouth.

I do not smile back.

'No.' I take another breath and my shoulders lift. They stay up high and tense as I breathe out through pursed lips. *Why on earth did I agree to this?* We move towards the door leading to the underside of the Sydney Harbour Bridge. I start nodding my head and telling myself I can do this. We move out on to a narrow metal grating and I can see through it to the ground. Some profanities leave my mouth and I grip tightly on to the metal bars on either side. I want to cry. Matt asks if I'm all right and tells me to keep moving forward. His words are like a match to a flame and I snap back at him.

'I *am* moving! Whose ****ing idea was this! I hate this!' I then realize I am still only on the underside of the bridge and it's only getting worse. When we get up on to the steps to climb the bridge, the muscles in my legs are shaking so violently that they hurt already. I'm vaguely aware that I'm making quiet noises that are somewhere between whimpering and groaning. I know there is no way back, so I keep putting one foot in front of the other. As we reach the top of the bridge, 134 metres high, the guide stops and turns.

'Why is he stopping? Why is he stopping?' More profanities under the breath.

He says something about the view and I'm not interested. He then asks us all to turn around and look behind us. I don't want to take either of my hands off the metal bars that I have been gripping, so I turn my body as much as I can without letting go.

That's when I see Matt kneeling down, holding out a ring box.

The tears were already there, ready to roll. I manage to let go of the bars for a split second so I can turn around fully, before gripping them again.

All the way through our beautiful moment my hands remain tightly gripped to the handrail.

The group applaud and start moving forward to cross the centre of the bridge and come down the other side. We take a moment to talk. I start asking about how he managed to arrange this. He explains as we walk across the centre of the bridge and scale back down the steps on the other side. I'm smiling and laughing and shaking my head. Matt explains that family who live here in Sydney, and his family members who came on the trip with us, are all watching from the restaurant opposite the steps we are descending on. I look across and see them all waving. I wave back with one hand and hold the ring up with the other.

Then I realize, I'm not holding on. I haven't been holding on the whole way down.

Our brains take in and process a lot of information every second of every day. But the world around us offers infinite amounts of information. If your brain was to try and process everything, you would not be able to function. So the brain makes choices about what to focus on. Our attention is like a spotlight. We have control of that spotlight, but we cannot control the actors who come on stage. We cannot control how long they spend there, what they say, or when they leave. What we get to do is focus that spotlight on one or two of them at a time. If we settle our focus on the anxious thoughts that tell stories of worst-case scenarios and images of you not coping, they get the chance to feed back to the brain that all is not well. When you shift the spotlight of your attention to other thoughts on the stage that offer a different story, they will have their influence on your bodily reaction too. While you are focusing on them, the other thoughts may not leave the stage. They may stick around, waiting for the spotlight again. But without it, they have less power over your emotional state.

The story of my engagement is rather an extreme example, but one that has stayed with me on the power of the focus of attention. All the way up the bridge, I was focused on how I might accidentally die that day. All the way down, I was focused on living.

Of course we can't rely on a surprise proposal to shift our attention away from our catastrophizing thoughts every day. But exercising the power to direct your spotlight of attention is a powerful tool. It is not the same as blocking thoughts out. As soon as you try to eliminate a thought from your mind and not have it at all, the thought pops up more than ever. This is how people get stuck in loops with intrusive thoughts. If you are not willing to have it, you will. By thinking about how you don't want to have those anxious thoughts, you are directing your spotlight on to them. When you choose to move the spotlight on to other thoughts, the anxious thoughts may remain on the stage. You are still aware of their presence, but they are not the star of the show.

When those anxious thoughts arise and you put the spotlight on them and start to ruminate over that feared event in the future, doing so causes your body to respond. Not only that, each time that you play out the worst-case scenario in your mind, of something awful happening and you not coping with it, you are constructing an experience that your brain uses to help build your concepts or templates for the world. The more you repeat that, the easier it becomes for your brain to re-create it.

Where you direct the spotlight of your attention helps to construct your experience. So learning to take control of that spotlight is truly an investment in your future emotional experience of the world.

So what if there are no other actors on the stage? How do we choose what else to think, when we are well practised at worrying?

What to focus on
instead – a new self-talk

Anxious thoughts are threat-focused. When we spend time with them, they feed back to your body and brain to ramp up the threat response. To turn the dial down on that threat response we instead need to cultivate a thought stream that promotes calm.

When my son had an operation at the age of two and a half, the swelling of his face caused his eyes to seal shut. He awoke from a nap unable to open his eyes. He could hear all the strange sounds of the high dependency unit. Lots of beeping machines, footsteps and voices he didn't recognize. His threat response fired up and he screamed for me. He was inconsolable until I arrived back in the room, held my hands on his and spoke to him. I could not make him see. I could not change his pain. I did not have any magic words to make it all go away. I simply spoke calmly in his ear, letting him know that I was there and he was safe. The person who had his back was here and going nowhere. From that moment on, his ability to accept and move through this frightening situation was nothing short of extraordinary. Over the next few days his eyes remained closed but he continued with life, playing with his toys and enjoying himself. Compassion had helped him to feel safe enough to face the world, even when all was not OK.

When we receive kindness and compassion, it turns the dial down on our threat response and allows us to feel safer. This happens whether that kindness comes from another person or from our own thoughts. Changing the way you speak to yourself changes your brain chemistry and your emotional state.

It is not easy. One day of self-compassion is not going to overturn a lifetime of practising self-criticism and self-attack. It is a life-practice that needs constant work. But it can be life-changing. Remember, compassion is not always the easy thing. It is not saying there is nothing to be scared of. It is the coach in your ear with a calm and firm voice that is encouraging you, supporting you, reminding you that you can and will move through this moment.

One of my favourite ways to turn my attention to a compassionate thought process is to ask myself, if I was coaching a friend through this, what would I say and how would I say it? The best kind of coach is not one who swoops in to rescue you, but one who is honest with you and encourages you to find the strength within yourself to move through difficult moments, so that you may discover your own strength.

Reframing

At the end of clinical training, I had to undergo a viva exam. This is an exam that is more like an interview, in which you sit in front of a panel of experts and answer questions on your research. On the day of my exam I arrived at the university and entered the waiting room where we had to sit before we were called in. As I sat there listening to my heart pound, a fellow trainee returned to the waiting room with tears streaming down her face. She was sobbing as a member of staff put an arm around her shoulders and led her out of the waiting room. All remaining pairs of eyes in the waiting room widened and looked at each other in terror. Something jumped in all our stomachs. I stood up and moved out of the room, passing a tutor who wished me luck. Then he gave me one of the best pieces of advice I ever received.

He told me to try and enjoy the exam. He told me that this was an opportunity to show off everything I had learned and worked on during these years of training. He said this was the one and only time that anyone would read the entirety of my thesis and show genuine interest in it, so it was my chance to enjoy sharing that interest. I returned to the waiting room nodding and smiling. What I didn't recognize until I was on safer ground was how he had helped me to reframe the entire experience. Nothing changed about the high-pressure situation I was facing. But I went from a rabbit in headlights to creating an experience that included a mix of courage, pleasure and excitement.

In the same way that my viva exam was reframed from a risk to a challenge, you can use this same technique with other experiences that you might otherwise interpret as a threat or something you can't cope with. Reframing does not mean that you deny the inherent risks in a given situation. There was still a risk of failing my exam. But if I chose to focus exclusively on that risk then my stress response might have been much higher and I probably would have found it much more difficult to perform.

Reframing is when you allow yourself to consider reinterpreting the situation in a way that is going to help you move through it. Reframing an experience as a challenge can help us to shift from the flight urge to a somewhat more controlled fight urge. We can move towards something with intention. The next step can help to put reframing into place.

Consider values and identity

When anxious thoughts are dominating the spotlight, we need to bring on stage the thoughts about what matters most to us. It makes sense to base some decisions on fear. If your life is at risk, that is when those thoughts are most valuable to us. But life becomes more rich and full when we make our decisions based on our values and what matters most.

An easy way to do this is to come back to questions such as 'Why is this so important to me? In a year from now, when I look back on this moment, what action or response would make me proud and grateful? What kind of person do I want to be in this situation? What do I want to stand for?'

Your values can become a part of your identity too. Whether you want to be an adventurous person or a fit and healthy person or a sociable and friendly person, being clear on that can help you bring forward an alternative set of thoughts to the anxious ones. If you feel anxious about starting a conversation but you have decided that you are going to see yourself as someone who is sociable and friendly, this helps to create a template, a concept for how you will behave in social situations, even when anxiety is whispering in your ear to avoid conversation. Or perhaps I choose my identity to be someone who lives with courage, then I can ask myself how I would respond to this situation. What would my next move be if it was based on courage? What response would make me proud to write in my journal tonight and look back on this time next year?

Chapter summary

- Get some distance from anxious thoughts by labelling the biases when you spot them.
- Remember that, even when anxious thoughts demand constant attention, you can control the spotlight of your attention.
- Kindness turns the dial down on our threat response, whether that kindness comes from someone else or inside your own head.
- Reframing a threat as a challenge invites us to be courageous.
- Act in line with your values so that you are basing your decisions on what matters most, not on fear.

CHAPTER 26

Fear of the inevitable

The fear of all fears is that of our own mortality. Every human lives with this one inevitability that life must come to an end and the ultimate uncertainty that we don't know exactly when or how it will happen. This fear of the known and the unknown constantly threatens to undermine our peace and contentment in the here and now. Even thinking about the prospect of our own death can make us feel instantly powerless and frightened, while at the same time triggering a sense of the meaninglessness of life.

For some, the fear of death invades everyday life in a direct way that causes us to worry about the prospect of death at every turn. For others it can bubble up in unexpected ways, disguised as other, seemingly smaller fears around health and risk-taking. Both have the potential to disrupt and even destroy our quality of life.

It has been argued that a fear of death underlies many other mental health problems (Iverach et al., 2014). Health anxiety fills us with fear of getting sick and going to hospital, with the possibility of dying in pain. Those who experience panic attacks commonly misinterpret the pounding heart for a heart attack and the terror of believing that they are about to die triggers a panic attack. Many specific phobias, whether it is of heights, snakes or blood, all centre around predictions that death is more possible when in contact with those things.

The prospect of death is constant throughout our lives, but we cannot live in that constant state of fear. So we protect ourselves by filling our lives with safety behaviours that protect us from its constant threat. We may put strict limitations on any risk-taking, we strive for concepts of immortality through fame or fortune, by our connections to others and how we wish to be remembered by them. And who can blame us? Irvin Yalom, Emeritus Professor of Psychiatry at Stanford University, describes it perfectly in his book, *Staring at the Sun*.

'It's not easy to live every moment wholly aware of death. It's like trying to stare the sun in the face: you can stand only so much of it.'

He also suggests that 'though the physicality of death destroys us, the idea of death saves us'. In this sense, our very human anxiety around death is not just a discomfort to be eliminated. Confronting our awareness of death can also become a profound tool for finding new meaning and purpose in how we live. The fact that we all die can define the meaning that we give to life and help us choose how we live with more careful intention. In the same respect, the meaning that we attach to death can impact on our wellbeing today (Neimeyer, 2005).

In my own research with breast cancer survivors, many of them reported positive life transformations that they attributed to being faced with death. The experience triggers a surge in fear while also inviting them to re-evaluate what meaning they were going to place on their finite time. Higher scores on trauma reactions were associated with more post-traumatic growth and positive life transformation.

But we don't have to come so close to death to face what it means to us. In Acceptance and Commitment Therapy (ACT) we can do this by exploring the idea of our own funeral or thinking about our own personal heroes who are no longer alive. These exercises invite us to consider living, not in spite of the fact that it will all end, but because of it. Coming face to face with the questions about what you want your life to stand for can open us up to both turmoil and transformation. While it can be painful, it is not about dwelling on that anguish but to empower choice. For example, you imagine that you have been able to live your life according to what matters most to you. Then you allow yourself to consider what that would look like. If you had lived with the meaning and purpose that you chose, how would you behave day to day? What would you work hard on? What would you let go? What would you commit to, even if you might not be able to complete it?

Exploring death in that way can help us get clear on what matters now.

Our fear of death seems impossible to eliminate because we

know it is going to happen. The fear is understandable and the prediction is realistic. But our unrealistic beliefs around death make a rational fear worse. Much worse, to the point that it can interfere with normal day-to-day life. Those beliefs might be something like, 'My family won't be able to cope without me,' or 'The pain of death will be torturous.'

If we talk about our fear of death, most people respond with an attempt to reduce our fear by challenging the probability that it will happen any time soon. This is often well-intentioned but not helpful, as we all know that it will happen eventually and we all know that it can happen without warning. When we try to avoid the fear of death by making ourselves feel safe from it right now, that same fear will inevitably pop up again somewhere else when we are reminded of life's fragility.

What we are looking for is the need for deep acceptance of the certainty of death as a part of life and the uncertainty of how it will come about. For some, those two facts of life are the source of meaning for life itself. For others, we try not to think about it and live as if it might not happen if we stay safe enough. We avoid anything to do with death. We avoid talking about it and seeing it. Those avoidance patterns build up around things that we perceive as risky and our estimates about the level of risk start to increase along with the level of our anxiety about it.

When this happens, multiple phobias can pop up through your life. But unless we address the fear of death, one phobia calms only for another one to pop up a little while later.

So what can we do if we are consumed with fear of the

worst and we know it will definitely happen? Ultimately, if we want to live fully without day-to-day life being disrupted by our fear of death, finding our own way towards acceptance about death as a part of life is something we must do. Acceptance is not saying that death is something we want. It is giving up the struggle against the parts of our reality that we cannot control.

Acceptance of death is not the same as giving up on life. Quite the opposite. Acceptance of death allows us to bring meaning to life. In turn, building a sense of meaning in life and getting to work on living in line with that can allow us to accept death as a part of life.

It can change the way we live. We can live in a way that is informed by our values and has meaning. We can pay more attention to those things that matter most to us and live with purpose.

The loss of someone we know and the resultant grief can put us in touch with our own mortality. If that person can die unexpectedly, then so can I. What does this mean for me and my life? What meaning does today have?

Changing our relationship with death

There are different ways that people cultivate a sense of acceptance around death. Three of these, listed below, were originally proposed by Gesser, Wong and Reker (1988).

- **Approach acceptance** – Holding beliefs about an afterlife or the possibility of going to some form of heaven enables the individual to cultivate acceptance of their own mortality.
- **Escape acceptance** – For those who experience great suffering in their life, death may be accepted or even embraced as it is perceived as possible relief or escape from that suffering.
- **Neutral acceptance** – This is when death is perceived as neither desirable nor a means of escape from suffering, but as a natural part of life that we have no control over.

 Try this: A task sometimes used in ACT is to imagine being able to write your own epitaph. If you were able to write just a few lines on your own gravestone, what would you most want it to say? This is not a guess at what others might say, but a way for you to explore what you want to stand for. The meaning you wish to live by, starting from today (Hayes, 2005).

For anyone who may struggle with these exercises, working with the support of a therapist is advisable.

Try exploring some of your own beliefs around death that make the fear worse. We each hold numerous beliefs around death that are either helpful to us or detrimental. One example might be that death is somehow unfair and we shouldn't have to go through it. Such a belief is likely to feed the anxiety and increase distress when thoughts of death arise. Exploring and taking time to challenge some of those beliefs is worthwhile. However, some of this work is so emotive that it helps to do it alongside someone you trust. That may be someone you know or a therapist who can help to guide you through it.

 Toolkit: Writing to unravel our fears around death

Expressive writing on the subject can help us explore fear of death because it allows us to pull away and ground ourselves without losing the thread of our insights and discoveries along the way. You can stop at any time and then return to the task when you are ready.

Facing a fear of death is not easy, and this is where a highly trained therapist can make a huge difference. For those who don't have access to such options, connecting with a trusted friend or loved one can also be a great support, as this is something we all face.

Here are some prompt questions you can use in journalling, therapy or supported conversation with a loved one.

- What are your fears about death? How do they show up in your day-to-day life?
- Which of your beliefs about death are different to others?
- What do these differences tell us?
- How have your past experiences of endings or loss shaped your beliefs about life and death so far?

- What behaviours do you engage in that help you feel safe from death?
- What would you like your life to mean or represent?
- What footprints would you like to leave after your time here?
- How can that meaning translate into real actions and choices that you make today and as you move forward into this next chapter of your life?
- Imagine that far into the future you were near the end of your life and looking back on this chapter that is just beginning. If you were to look back with a smile on your face, feeling content and satisfied with the choices you made and the way you approached each day, what would daily life need to look like?
- If this next chapter of your life was to become the most meaningful and purposeful, what would it include?
- If your awareness of death was to enhance your life rather than diminish it, what would that look like?

Chapter summary

- Our collective fear of death is a fear of both the known and the unknown.
- For some people, coming close to death brings about growth and positive life transformation.
- Acceptance of death is not the same as giving up on life. Quite the opposite.
- Acceptance of death allows us to bring meaning to life.

7

On Stress

CHAPTER 27

Is stress different from anxiety?

Stress and anxiety are both terms that have become widely used as umbrella terms for a diverse set of experiences. It is not unusual to hear people say that they were stressed and so their anxiety got worse. Or they might say the opposite. The result is that most people use the words interchangeably to describe an almost infinite number of experiences. They might be stressed about work deadlines or anxious about finding a spider in the bathroom. You might feel stressed because you had to wait in a queue at the post office that made you late. But you might also describe dealing with stress at losing your job and struggling to pay rent. The next person might describe both of those as anxiety-provoking.

But you'll notice that the two are each given their own

sections in this book. The experience of what we call stress is constructed through the same mechanisms in the brain as emotions are (Feldman Barrett, 2017). Your brain is constantly receiving information from your body about the demands of the outside world and trying to work out how much effort is needed. It tries to match the amount of energy being released in the body to the demands from the outside world, to ensure that nothing is wasted. When our internal physiological state feels well matched to the environment, we mostly interpret that as a positive feeling, even when it involves stress, for example, when you feel pumped and ready for a big sports competition. But when that internal environment is not matched to the demands of the external world, we tend to interpret that as negative. When we are tired but wired and can't get to sleep. Or when we are so stressed that we can't seem to focus on the questions in an exam or job interview. At such times we tend to experience the prediction that we cannot cope with current demands.

Stress and anxiety are both associated with states of alertness. But for the purposes of this book, anxiety is associated with the feeling of fear and the excessive worry thoughts that come along with that experience. In contrast, the stress that you feel in the line at the post office would have a different meaning to that of anxiety. If you felt stressed in the line, that might be because you had a tight schedule of things to get done that day. That spike in stress would increase your alertness to help you make a decision about whether to continue in the line or re-prioritize in order to meet your expectations of the day.

If the feeling was anxiety, it is likely to be associated with worry thoughts and predictions that something threatening or dangerous might happen.

So while the mechanisms of stress and anxiety are the same, we conceptualize them in different ways. If you are lying in bed and hear glass breaking downstairs, your stress response will fire up but you are more likely to frame that sensation as anxiety and fear. You might have the urge to fight off the threat or escape. That same stress response is given different meaning when you are carrying around the threat of unemployment or struggling to manage the demands of both a job and parenting. Those things are not perceived as immediate danger. We cannot fight them or escape them in the same way.

So while we have simplified that stress response down to a fight or flight, in reality there are many ways in which the stress response can vary. There can be differences in the ratio of hormones released, cardiovascular changes and other physiological responses that combine to form different psychological experiences and different behavioural urges.

We feel stress when our brain is preparing us to do something. Whether that thing is to get up in the morning, start a work presentation or drive your car, your brain gives out energy to increase alertness and ensure you are ready to react to your environment, whatever that may be. Cortisol, which we all think of as the damaging stress hormone, is actually enabling the quick release of energy in the form of glucose into your bloodstream for fuel. Your lungs and heart start working faster to deliver the necessary energy from oxygen and sugar to the major muscles and the brain. The adrenaline and cortisol then help your muscles to make the most efficient use of that energy. You are geared up ready to face whatever challenges you are presented with. This is your body working at its best. Your senses sharpen and your brain is processing information more rapidly.

When your brain gives out those resources, it expects to get something back for it in terms of rest or nutrients. But when it does not get anything back, there is a shortfall. If that happens repeatedly, your body's resources are not replenished. If you're not sleeping enough or eating well, or if you are arguing with your spouse every day, the shortfall gets bigger. Over time that depleted body struggles to defend itself and becomes vulnerable to illness.

If you are faced with a threat to your survival, the experience will be the fight or flight response. But when you perceive a stressful situation that is not such an immediate threat, you may experience more of a *challenge response* which allows you to rise to the challenge in much the same way, but the sensations feel less like intense fear and more like agitation to move.

Anticipatory stress is experienced when we can predict that something stressful is coming up and will demand a lot from us. You know you are going to feel nervous and stressed in that job interview next week, so you start anticipating that challenge ahead of time. When we get it wrong and keep predicting that we will face a challenge that we cannot cope with, we become anxious as we start to fear the physiological and psychological discomfort of that stress. When the stress is triggered by a physical threat and the body is activated to move, the process of moving and reaching safety invites the body to come back to baseline. But if we keep triggering the stress response for those psychological reasons, then the physiological upheaval is not so short term and there is no clear path for bringing back calm. That is when we start to get into trouble with effects on physical health, mental health and behaviour (Sapolsky, 2017).

Chapter summary

- The terms stress and anxiety are often used interchangeably.
- When we are able to meet the demands of our environment we tend to feel good, even when that involves stress.
- What we feel as stress is when our brain is preparing us to do something.
- The brain allows the release of energy to increase your alertness and react to your environment.
- We conceptualize anxiety as a more fear-based response. But this is one way the overall stress response can vary to meet your needs.

CHAPTER 28

Why reducing stress is not the only answer

Reducing stress where we can is generally a good idea. But reducing the stress in your life is so frequently offered as a stress management solution and it has never sat comfortably with me. One reason is that it is a vague term that nobody really knows what to do with, and the second is because many stressors are non-negotiable.

While some stress in our lives is stress that we choose (the stress leading up to a sports competition or preparing for a big event like a wedding), the most intense stress we face is often not optional. A high-pressure situation might be a boxer entering the ring but it might also be entering the hospital to find out the results of your biopsy. It is working through your finances and realizing you might lose your home. These are the moments that bring huge stress responses and need real-time tools to help us deal with them in the healthiest, most efficient way possible.

We humans have a love-hate relationship with stress. We love the thrill of a horror movie or the speed of the roller-coaster. We actively choose these spikes in our stress response and we anticipate them with huge excitement. We may feel out of control, but we know it's just for a moment. We feel afraid, but at the same time we trust that we will live to tell the tale. We maintain enough control to stop the experience at any time. Too little stress and life is boring. Just enough and life is engaging, fun and challenging. Too much and all of those benefits can be lost (Sapolsky, 2017). We need a fine balance between predictability and adventure.

In just the same way that emotions are not all bad, stress is not all bad either. It is not a malfunction or weakness of our brain or body. It is a series of signals that we can use to help us understand what we need.

Stress has positive effects in the short term. The release of adrenaline in the stress response helps to fight both bacterial and viral infection in the body. It increases heart rate, sharpens cognitive function and dilates the pupils. All of this helps you to narrow your focus, evaluate your environment and respond to meet its demands on you.

We are led to believe by the popular view that stress is an outdated survival mechanism that is no longer necessary. This means that when we start to feel its effects, the pounding heart and the sweaty palms, we believe we're failing to cope or the body is letting us down. It is seen as a fault in the system or a signal of disorder that needs to be shut down. But the story is not so black and white. Stress is not always harmful and our main aim does not always need to be to eliminate it.

Science has taught us about the dangers of stress but has also revealed a more complete story about its function, how we can use it to our advantage, and how best to replenish our mind and body to stop stress from becoming dangerous.

So when you feel those signs of stress as you start a presentation at work or in school, your body is helping you to perform at your best. In those situations, we don't want complete calm and relaxation. We want to be alert and clear-thinking so that we can achieve whatever goal we are working on at the time. What we don't want is for that stress to be so high that it has a detrimental effect on our performance or causes us to escape and avoid. Learning how to turn the dial down on that stress when we don't need it and turn it up when we do need it is the foundation of healthy stress management.

We cannot untangle stress from a meaningful life. Whatever your unique personal values, anything that you strive towards and work for is going to require your stress response to get you there. The stress response is a major tool in enabling us to reach our goals. It is often the things that are most important to us that have the potential to bring most stress. If it matters, it is worth moving for. So experiencing stress is not just an indication of problems or warning signs for health issues. It can also reflect a life in which we are acting on things that we care greatly about and living a life with purpose and meaning. If we can learn how to use it to our advantage, and turn down the intensity when we need to, then stress can be our most valuable tool.

Chapter summary

- Stress is not always the enemy. It is also our most valuable tool.
- Learning to replenish after a period of stress is more realistic than trying to eliminate it.
- Stress helps you to perform and do what matters but we were not built to be in a constant state of stress.
- We need stress for a fun and challenging life, but too much and the benefits are lost.

CHAPTER 29

When good stress goes bad

The stress response works at its best when it is short term and limited. When our circumstances cause continued stress that we cannot change, or we don't know how to bring that stress back down, our body is not replenished for its effort. Imagine driving along the motorway at speed in second gear. There is only so long you can keep that up before damage is done.

When stress becomes sustained over long periods, our brain tends towards more habitual behaviours that demand less energy. Our ability to control our impulses, remember information and make decisions becomes impaired. Over time, our immune system is affected. In the short term, adrenaline gives our immune function a boost to help fight bacterial and viral infections. But in the longer term, over-production of adrenaline and abnormal patterns of cortisol are linked with shorter life expectancy (Kumari et al., 2011). When adrenaline is repeatedly propping up our immune system through chronic stress and then we stop and the adrenaline goes down, so does the immune system. This is why you often hear of people who work incredibly hard around the clock for months on end and when they finally stop to take a holiday they almost immediately fall ill.

Burnout is a term used to describe the response to excessive and prolonged stress at work, although paid employment is not the only environment in which we can experience burnout. Anyone in a caring role, parenting role or volunteering role may also experience burnout.

People often describe feeling emotionally exhausted and drained, as if they are running on fumes and have no resources left. They may notice feeling detached from other people, or themselves. They often report feeling that they lack competence at work or at home and no longer have that same feeling of accomplishment they once had from those things.

Burnout happens when that short-term stress response that we have is repeatedly triggered over a prolonged period, without enough chance to rest and restore in between. There is often a chronic mismatch between the individual and one of the following:

1. **Control** – Living in a situation in which you do not have the resources needed to meet the demands you are faced with.
2. **Reward** – This might be financial in an employment scenario. But equally, it can be a sense of social recognition or acknowledgement of value, either in a work environment or any other.

3. **Community** – A lack of positive human interaction and the sense that one has social support or a sense of belonging.
4. **Fairness** – When there is perceived inequality in any of the other factors in this list. When some people have their needs met more than others or demands fall on some more than others.
5. **Values** – When the demands you face are in direct conflict with your personal values.

Let's be clear. Burnout is a serious health issue. Anyone who thinks they may be experiencing burnout needs to act as soon as possible. But we also have to be realistic. Some pressures you can say no to (like taking on that side hustle on top of a 50-hour working week) but others you can't (such as physical illness or financial pressures or the emotional strain of a relationship breakdown).

When you are fighting to keep a roof over your head and feed your children every day by working two jobs, and trying to be the best parent you can in between, there is no option to just remove stress from your life and take on an idyllic morning routine of meditation and yoga. But addressing burnout does not need to look like a holiday photo. Living with high demands and the associated stress while looking after your health is a balancing act in which you will sway from one direction to the other. There is no silver bullet that will fix everything. What works to help one person balance demands will not be realistic for the next.

When we are not able to turn that dial down on stress, or we are overloaded for too long, that stress can become chronic. The signs of chronic stress differ between individuals, but I have listed a few below.

Signs of chronic long-term stress:

- Disturbed sleep on a regular basis.
- Changes in appetite.
- More frequent agitation and irritability that may impact on relationships.
- Problems concentrating and focusing on tasks.
- Problems switching off and resting even when exhausted.
- Persistent headaches or dizziness.
- Muscle pain and tension.
- Stomach problems.
- Sexual problems.
- Increased dependence on addictive behaviours such as smoking, drinking or over-eating.
- Feeling overwhelmed and avoiding small stressors that would normally feel manageable.

 Try this: If you think you may be experiencing burn-out, try answering the following questions. Then spend some time reflecting on your answers and what this means for you. There are validated measures of burnout (Kristensen et al., 2005 and Maslach et al., 1996) but you are the expert of your own experience.

Reflecting on how your current situation is influencing your health can help you to acknowledge when things need to change.

- How often do you feel emotionally drained?
- When you wake up in the morning, do you feel exhausted at the thought of everything you have to do?
- When you do have free time, are you left with enough energy to enjoy that time?
- Do you feel persistently susceptible to physical illness?
- Do you feel able to deal with problems as and when they arise?
- Do you feel that your efforts and accomplishments are worthwhile?

The communication between brain and body goes both ways. This means that when your body is under stress for extended periods, the persistent messages about this make changes to your very adaptable brain that is trying to regulate your body. This is why stress is so very damaging to both physical and mental health. It affects all aspects and every part of you (McEwen & Gianaros, 2010).

In the balancing act of managing stress and using it to our advantage while remaining healthy, we need to balance incoming demands with replenishment. The more demands on us, the more replenishment we need. The more stress pouring into the

When good stress goes bad

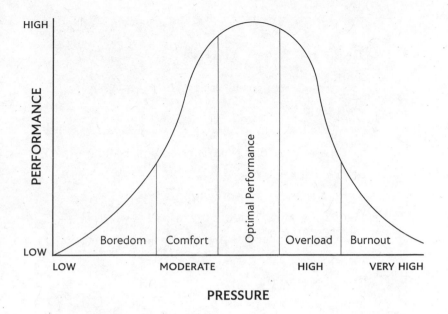

Figure 9: The stress curve. A certain degree of stress will help you perform at your best. Go beyond that, and your performance declines.

bucket, the more release valves we need to process it and make room for the ongoing demands.

The good news is that we can reduce the effects of stress on the body with a few simple tools that I will describe in the next chapter.

Chapter summary

- The stress response works at its best when it is short term.
- Chronic stress is like trying to drive your car on the motorway in second gear. There is only so long before damage is done.
- Burnout is not just a workplace issue.
- There is no silver bullet that works for everything. The right balance for one person will be unrealistic for another.
- If you are showing signs of burnout, listen to them and respond now by starting to meet your needs.

CHAPTER 30

Making stress work for you

In the previous section on fear I talked about using breathing techniques as a fast way to calm the body and mind (see pages 200–202). Those techniques are just as helpful with stress. Your breathing can directly impact on your heart rate and level of stress or calm. When you inhale, the diaphragm moves down, creating more space in the chest which allows the heart to expand further and the rate of the blood flow to slow down. When the brain receives the information about this, its job is to then send a signal to speed up the heart.

In contrast when you breathe out, the diaphragm moves up leaving a smaller space for the heart and so the blood passes through quicker. This causes your brain to send signals to slow your heart down.

- When outbreaths are longer and more forceful than inbreaths, this slows the heart rate and calms the body.

- When the inbreaths are longer than the outbreaths, we become more alert and activated.

Therefore, one of the most immediate ways to start calming your stress response is to make your exhales longer and more forceful than the inbreaths.

It's worth noting that the aim when you are feeling overwhelmed with stress is not to go from agitated and worried to a relaxed and meditative state. When the world is demanding of us we want to be alert. As you use breathing techniques such as this one, you will notice that your mind's ability to think more clearly and problem-solve becomes more available to you. In that sense, we are not trying to make it all go away and induce ultimate relaxation, but to put you in the best state possible in which you are able to use the advantages of the stress response (e.g. alertness) and bring down the intensity of the disadvantages (e.g. worry and overwhelm).

However, if you are putting time aside to practise relaxation or have interest in using breath work, then you can use this technique over longer periods to bring about the physical state of deep relaxation. That tends to be when you have more time and no distractions or current demands, so that you can use longer outbreaths or other relaxation techniques to help you relax. But when you are in the moment and you need to perform, then the breathing technique is a good choice for helping you get through the moment.

Turning to others

I'm sure most parents can relate to the experience of lying in bed and running through in their minds what they'd do in the event of a house fire. You run through each possible scenario of how you could grab each of your children as quickly as possible. How does that need to protect fit into a fight or flight mode? Fight or flight does not tell the whole story. Connection with and protection of others is just as much a part of our survival instinct as fighting that fire or escaping from it. Some stressful situations may lead to more selfish behaviour, but other scenarios lead us to be even more caring towards others.

Research also shows that when we do focus on caring for others in times of stress it changes our brain chemistry in such a way that produces feelings of both hope and courage (Inagaki et al., 2012). It even helps to protect us from the harmful effects of long-term chronic stress and trauma. So it becomes a source of resilience (McGonigal, 2012). This tend-and-befriend stress response may have evolved for the protection of offspring, but the stress response is generic, which means we can apply those same feelings of courage to any scenario that you encounter. Connecting with others helps us recover from stress.

Social isolation in itself places the mind and body under great stress. Greeting people we love in person and engaging fully in our relationships mitigates the effects of short-term and long-term stress.

Goals

So much of what we are exposed to, especially in the self-improvement industry, is about being your best, standing out from the crowd and being exceptional. One of the first questions you hear when you meet new people is 'What do you do?' A fair question, but one that reflects our focus on careers. Life goals are often set from a competitive standpoint, everyone striving to prove themselves as enough with symbols of achievement. We are led to believe that happiness is somewhere beyond becoming exceptional. Many people who discover differently have had to learn the hard way through burnout and mental health crises.

But science is beginning to unravel this fallacy. Those who build their life on self-focused goals are more vulnerable to depression, anxiety and loneliness. Whereas those who structure their goals on something bigger than the self tend to feel more hopeful, grateful, inspired, excited and experience better wellbeing and life satisfaction (Crocker et al., 2009). Of course, we all have times when we focus on the self and times when we focus on goals that are bigger than the self. We have the ability to shift between these mindsets, which is crucial. We only have to spend a small amount of time reflecting on how our choices and efforts might help a greater cause in order to shift our experience of stress. When we focus on how our actions, big or small, can help others, we show less stress response in difficult and demanding situations (Abelson et al., 2014).

So what does this mean in the real world? When we make a conscious effort during stressful events to link our engagement to our values and to make a difference to others, we find the stress easier to cope with. We change the meaning of our struggle in such a way that we are motivated to persevere rather than to escape and avoid the stress. The test becomes less threatening because it is no longer about proving our self-worth. Our self-worth becomes inherent in our efforts to make a difference.

 Try this: How to shift from self-focused to something bigger.

When you feel under stress and you notice the urge to escape or avoid, take some time to return to a values check-in. Ask yourself questions like:

- How does this effort or goal fit with my values?
- What kind of contribution do I want to make?
- What difference do I want to make to others with what I am doing?
- What do I want to stand for during this? What do my efforts mean to me?

 Toolkit: Using meditation for stress

Meditation is not a belief system or a new-age fad. It is a technique that, as science is discovering, has a powerful effect on the brain and on our quality of life. Scientists continue to reveal more details of the process, but what is known is that it changes the structure and function of the brain in ways that help us to reduce stress and improve our ability to regulate emotion.

When we are under stress, those tend to be the periods when we have even less time to rest. Yoga nidra is a meditation technique that promotes deep rest and relaxation. It is a simple technique which is often carried out using guided meditation audio that leads you through perception exercises (e.g. focusing on the breath and areas of the body). It has been increasingly examined by researchers in recent years and shown to reduce stress (Borchardt et al., 2012), improve sleep (Amita et al., 2009) and increase general wellbeing. Most guided meditations last for 30 minutes, but recent research on 11-minute meditations have also shown that even short doses of yoga nidra can help with stress for those who are not able to meditate for longer periods (Moszeik et al., 2020).

So when demands are high and time is short, using short breaks to make use of yoga nidra would be a better option than 10 minutes of scrolling on social media.

Meditation is not a cure-all. Just like exercise, it is another potentially powerful tool in the box. There are lots of different types of meditation but here are some of the practices that have been scrutinized by the research:

- Mindfulness meditation. This is most widely promoted and is taught as a part of several psychological therapeutic approaches. It teaches the mental skill of staying alert to the present moment and observing sensations without judgement and without getting caught up in them. It is a great tool for using in the moment to help us deal with stress and emotions we are experiencing. It builds your ability to pull your mind back from thoughts of the past or future and to observe experiences as separate from the judgements and meaning that we attach to them.

- Meditations that use pictures or mantras (a word or phrase that has meaning or significance for you) or objects to help you focus your attention.
- Guided meditations that help us to cultivate compassion and kindness.

Being mindful does not mean you have to surround yourself with candles and meditate all day. Being mindful is about the practice of paying attention to the present moment and observing sensations as they come and go, without getting caught up in those sensations or struggling against them. It means staying open and curious to experience without judgement and without rushing in to attach meaning to them. Meditation allows us to practise mindfulness formally. As we learn to drive through driving lessons until the practice becomes instinctive, so the same is true of learning to practise mindfulness through meditation.

So whether you already meditate and you want to bring those skills into your day-to-day activity, or you struggle with meditation but want to have a go at being more mindful, here are a few ways you can put mindfulness into action:

Mindful walking

- Start by noticing the sensation of the soles of your feet. How it feels when they make contact with the ground. The movement of the foot as it lifts off the ground and moves forward. How much time it spends in contact with the ground.
- Notice the movements of your arms as you move. Not trying to change that, just noticing.
- Expand your awareness to notice the whole body and how it feels to be propelling yourself forward. Notice which parts of your body need to move to aid this process and which parts remain still to accommodate that process.
- Expand the attention further to focus on the sounds around you. Try to acknowledge the sounds that you might normally not notice, each time observing from a non-judgemental stance.
- Each time that your mind wanders and starts telling a new story, gently bring your attention back to your experience of walking in this moment.
- Notice everything you can see as you walk. The colours, lines, textures and the movement of your visual perception as you pass.
- As you breathe, focus your attention on the temperature of the air and any scent or lack of.

Mindful showering

For many of us, the morning shower is the time our mind gets to work planning the day, worrying about everything we have to do or dreading the moment we must leave the hot water and start our day. But in those few minutes, this can be a great opportunity for practising mindfulness. There is a lot of sensory information that feels unusual compared to the rest of the day, and so some people find it easier to anchor themselves in the present while in the shower.

- Focus your attention on the sensation of the water hitting the body. Where it hits you first, and which parts of the body are not in contact with the water.
- Notice the temperature of the water.
- Acknowledge any scents from soaps and shampoos.
- Close your eyes and listen to the sounds.
- Notice the steam and water droplets in the air or as they land on different surfaces.
- Notice any sensations in the body as you stand there.

Mindful teeth brushing

- Bring your attention to the taste.
- Acknowledge the sensation of the toothbrush as it moves.

- Notice the movement of your hand and the tightness of the grip with which you hold the toothbrush.
- Listen to the sounds of the brushing and running water.
- Notice the sensations as you rinse your mouth.
- Each time that your mind wanders, gently guide your attention back to the various sensations of this process that is happening right now.
- Try to notice this activity that you do every day with the same curiosity as something that is brand new to you.

You can do the same with any daily activity, whether it's swimming, running, drinking coffee, folding laundry or washing up. Simply pick a normal everyday activity and follow the prompts to engage with that activity mindfully.

Remember, if you notice your mind keeps wandering, you are not getting it wrong. Every mind is constantly wandering and making sense of the world. Mindfulness is not ultimate, unbroken concentration. It is the process of noticing when your mind shifts its focus and intentionally choosing to redirect that focus back to the present moment.

Awe

In the way that meditation can help us to get some distance from our thoughts and emotions, there is another experience that seems to have a similar effect. Awe is the feeling of being faced with something vast and beyond our current understanding of things. We can experience awe in the presence of beauty, the natural world and exceptional ability. Those moments that force us to re-evaluate and re-think things in order to accommodate this new experience. From coming face to face with a powerful and charismatic leader, to gazing at the night sky and contemplating the universe and the chances of having even been born. Some experiences of awe come in once-in-a-lifetime experiences, like witnessing the birth of a child. Others can be connected with more frequently. Perhaps a walk through the woods, looking out to sea or listening to a powerful singer.

Psychology research has neglected this area so far, but we see awe used by people to detach from the messy details of everyday life and to widen the focus from the small stuff to the wider world and something that feels vastly bigger. But since the birth of the field of positive psychology, the research is starting to acknowledge the importance of positive emotions and not just the eradication of the negative ones (Frederickson, 2003).

There is some relationship here between awe and gratitude but so far any empirical evidence is lacking. In hearing people talk about their experiences of awe, I hear people talking about feeling small, and in doing so, being able to recognize more easily what matters most. It appears to bring about gratitude and a wonder for having the chance to be alive. And it doesn't require you to live on a beach in Thailand or have access to Niagara Falls. It can be felt with a focus on ideas and imagery. Many self-help gurus and motivational speakers talk about how the chances of being born are one in 400 trillion. This idea is hard to even comprehend and forces us to spend time trying to acknowledge how fortunate we must be to have this chance to live, even for a short time. Those ideas trigger a sense of awe and the something-bigger-than-me feeling. There is nothing like feeling small in a vast universe to bring your stress down and feel comforted by the new perspective. In trying to accommodate that in your mind, everything needs a reshuffle and you get a slightly fresh perspective on whatever you feel consumed by.

So, when dealing with stress, why not explore what triggers a sense of awe for you, whether it is time with animals or nature, watching extraordinary performances or looking up at the stars. It helps to document these experiences, perhaps through journalling, so that you can understand the effect on you and then later return to memories even if you cannot return to the place.

Chapter summary

- Changing something as simple as how you breathe will impact on your stress levels.
- Science is showing us that meditation has a significant effect on the brain and how we deal with stress.
- Connecting with others helps us to recover from stress. Social isolation puts the mind and body under great stress.
- Goals founded in contribution rather than competition helps us to stay motivated and persevere under stress.
- Seek out experiences of awe to shift your perspective.

CHAPTER 31

Coping when it counts

We are so bombarded with information about how stress is bad that most interventions focus on getting rid of stressors and adding in more rest and relaxation. But what does that mean for the non-negotiable events that cause stress? How do you face the surge in stress as you walk into a job interview or exam? How do we deal with those moments and stay on our game? When you are faced with a high-pressure situation, all the wonderful research on how to relax and de-stress starts to feel less helpful. You cannot rush off to engage in a deep relaxation exercise as your exam begins, and vowing to be less perfectionist as you enter the only job interview you've managed to get in months is not going to make you any less stressed. In these situations, what we really need are some clear tools for how to use the stress to help us perform and even learn from that experience. We need to know how to actively cope with the demands of those non-negotiable high-pressure situations.

If there are times when stress works to our advantage, it is in these high-pressure, short-term situations and so the goal is not to eliminate that stress and stroll into the interview as relaxed as you feel at home on the sofa. Instead, the key is to use the benefits of the pressure without letting it overwhelm you and have a detrimental effect on your performance.

Mindset – your new relationship with stress

Research shows that how we think about our stress affects how we perform under pressure. A shift from perceiving the stress response as a problem to viewing it as an asset that will help frees up those individuals to spend less energy trying to squash the feelings and instead focus on meeting whatever demands they face. As a result, they tend to feel less worried about the stress, feel more confident and perform better. This mindset shift can be the subtle difference between 'Despite how stressful it is, try your best' to 'When you feel the signs of stress channel that energy and enhanced focus to do your best.' There is also evidence that doing this helps us to feel less exhausted by the stress (Strack & Esteves, 2014).

When we focus our efforts purely on reducing stress in the build-up to a big event, whatever it may be, we are reinforcing the misconception that stress is a problem to be solved. When you are trying to get somewhere and stress rears its head, take it with you. Let it help you to focus and energize and move with accuracy. You are built to perform under pressure and that is exactly what you will do. Reminding yourself of this changes the meaning of those signs of stress that might otherwise be seen as 'symptoms' of a problem. In fact, research shows that simply reminding someone that their performance improves under pressure improves their actual performance by 33 per cent (Jamieson et al., 2018).

Bad language

One way to change our mindset is in our use of language. The words we use can powerfully determine the meaning of a situation and our approach to it. Imagine you were a professional athlete and just before you left the locker room to compete your coach said, 'You're about to blow it.' Not only would your stress increase but your subsequent thoughts are likely to sound like catastrophizing thoughts that turn that stress into something that feels like panic.

Social media is awash with daily affirmations and quotes, some of which may happen to hit home if they reach the right person at the right time. But what difference can they make?

Some of them focus on what to stop doing or sweeping statements about what to avoid in life. There is a big problem with focusing an affirmation on what *not* to do. We only have the one spotlight of attention and when we focus it on what not to do, that leaves little room to focus on what we *do* need to do for things to go well.

Others focus on being purely positive. These can be uplifting but only if you believe in them. Simply being told to 'Stay positive' or 'You're doing great' is vague at best and offers no clear guidance on how to move through the challenge ahead of you.

Dr Dave Alred is an elite performance coach who works with many of the world's top athletes, helping them to perform at their highest level under extreme and very public pressure. When he puts together affirmations for his athletes he makes sure those affirmations steer away from sweeping absolutes and remain concrete and factual – something the player believes. They clearly identify the key to the necessary mindset and remind them that keeping to the process will lead to improvement. When statements make it clear what needs to be focused on, they give us direction. Alred (2016) suggests starting with a 'how to' statement, then vividly describing what happens when the process is right, and thirdly conjuring up the emotional state that matches your intention. When stress is high and threatens to disrupt our concentration or ability to perform, we can prepare statements like those in advance to help us match our thoughts, feelings and actions to our intentions. The type of demands you face will change the type of statements you need. The key is to keep them short, concrete, specific and instructional and for them to put you in touch with the feeling of the process that you may have practised before the event.

Reframing

We have covered reframing in other sections of this book but it is also especially helpful here. Reframing is using the power of language or imagery to adjust the way you perceive a situation. You are not trying to convince yourself of anything that you don't believe could be true. You are simply trying to shift your frame of reference. Looking at things from a new perspective can enable us to draw new meaning from the experience and, in doing so, shift our emotional state. In the section on fear we talked about reframing anxiety to excitement. In this case, we can reframe the sensations of stress to the feeling of determin-ation, or threat into challenge. Changing just these single words transforms that meaning without telling any lies about the real-ity of the physical sensations we face. With the latter words, we are choosing to embrace those feelings. With the former, we find them aversive and push them away.

Focus

In high-stress situations we tend to get tunnel vision. This is sup-posed to happen, as it helps us to focus on the most important demands. But if the sensations of that stress feel overwhelming there is something that we can do to allow the body to maintain that high output while calming the mind. Ongoing research into this suggests that choosing to shift from that tunnel vision back to a more panoramic view calms the mind. This does not mean moving your head to look all around you but just allowing your gaze to widen and take in more of your surroundings. The visual system is part of the autonomic nervous system and so dilating your gaze in this way accesses circuits in the brain associated with stress and levels of alertness. Huberman (2021) describes how this is a powerful technique for getting more comfortable with higher levels of activation. We don't want the stress response to stop because we often need it in high-pressure situations. We just want our mind to be more OK with it, raising our stress threshold.

Failure

When the pressure is on, it's often because the stakes are high. We believe that failure has big implications. This makes sense. When failure is interpreted as a big threat, the brain wants to focus on that threat to be sure we avoid it. For those who tend to self-attack after any failure, big or small, any signs of potential failure will likely lead to a spike in the stress response.

We all have a limited capacity for attention, and when we need to perform under stressful conditions we need to take full control of that spotlight and focus on what is going to help us face the challenge. To overcome that fear of failure in the moment and that preoccupation with everything that could go wrong, we need to immerse ourselves in a narrow focus on the process, leaving no space for worrying thoughts about potential outcomes.

This is where it can help to practise ahead of time, if it is possible given the situation. In building a familiarity with the process and how it feels to walk through it, you can pre-prepare those guiding statements that will remind you of what to focus on and expect if you need that on the day. If your process becomes a well-trodden path, you get the chance to build trust in that process.

Depending on the challenges that we face and the realities of failure, we can use the same skill of reframing to change our perception of failure.

 Try this: If you want to explore this subject with journalling, here are a few prompts to help guide you.

- How do you respond to your own failures?
- Do you deny them and quickly move on, forgetting it ever happened?
- Do you immediately start on the self-attack, name-calling and blaming something in your character?
- Or do you look outward and start blaming the world for making life so hard for you? If there is something that we don't get taught about enough, it is how to cope with failure.

When we believe that mistakes and setbacks are linked to who we are as a person and our self-worth, then even the smallest of failures will trigger shame and the urge to give up, withdraw, hide away and block out the excruciating feelings. This happens a lot for perfectionists. There is a focus on being enough in the eyes of others and assuming that those others demand nothing less than perfection. If I fail then I am a failure. If I lose then I am a loser. However small and temporary that setback may have been.

But when we respond to failure without these global attacks on our personality and instead focus on the specifics of the moment, holding our awareness that imperfection is an intrinsic part of our common humanity, the emotional result is different. Feeling guilty about an error in judgement or a choice made allows us to be honest with ourselves about where we went wrong without feeling doomed to being a failure forever. It focuses on the specific behaviour rather than attacking us as a person.

Crucially, you still take accountability for your actions. Self-compassion is not letting yourself off the hook constantly. It is focusing on the specific mistake as an isolated event so that you are free to learn from it and shift direction back towards your values. This is the path to continuing to improve and moving on from mistakes. Shame, on the other hand, immobilizes and paralyses us.

Failure is always difficult and heightens our stress response. In times of stress our negative core beliefs can become activated (Osmo et al., 2018). We start to entertain thoughts like, 'I am a loser, I am a complete failure, I am worthless, I am nothing.' Those thoughts and the shame that accompanies them are very powerful in making us feel completely alone and isolated. We buy into those thoughts as facts. We think we are the only ones and so we hide how we feel. But as it turns out, among the

7 billion people on this earth, these sorts of core beliefs are part of a list of just fifteen or twenty common negative core beliefs that are seen across the world. Essentially this means that we are the opposite of alone. As human beings, the need to feel worthy of love and to have a safe group in which we belong is in us all.

When we feel shame around failure we can feel as if our acceptance and therefore our survival is under threat. It is an all-consuming sensation that can stop us even trying to fix things because we believe that the problem is us, rather than a specific behaviour or choice.

When we are going out into the world and taking risks, making ourselves vulnerable to shame, we need the skills to manage that shame and move through it. We all need a safe place to return to that allows us to learn from failure without our worthiness as a human being coming into question. That place has to be our own mind. When someone we love is suffering, we show them kindness because we know it is what they need. When we take a fall, it is time to do that for ourselves. It is the surest way of ensuring we get back up and move forward.

But how do we become less hostile to ourselves and become instead the voice we need to hear?

Shame resilience

When we feel shame in response to a failure there is often a huge thought bias involved. We take one event, action, choice, or even a pattern of behaviour and we use it to make a global statement about who we are and our worth as a person. This makes a judgement on the whole person using only this specific information, neglecting every other layer of your strengths, weaknesses and intentions. This is something we wouldn't do to someone we love. If someone you love unconditionally made a mistake, you wouldn't want them to write themselves off as a person. You would want them to learn from the experience and move forward, making choices that were more in line with who they wanted to be. You would still want the best for them and so you would not subject them to a barrage of verbal abuse.

 Toolkit: Building shame resilience

Shame can be intense and extremely painful. Here is a list of tips for building resilience to the shame associated with setbacks and failure:

- Be vigilant about your choice of language. 'I am . . .' statements lead into those global attacks on character and your worth as a person that feed and re-trigger the shame.
- When reflecting on what happened, stay very specific to the behaviour that you considered to be a mistake. One behaviour or set of behaviours is not the whole you.
- Acknowledge that you are not alone in feeling this way. After a failure or setback, it is normal for most humans to

become vulnerable to feelings of shame and to focus on thoughts of self-loathing. Those thoughts are seen across the world but are not necessarily helpful or accurate.

- Acknowledge that this feeling, while painful and intense, is also temporary. We can use self-soothing skills (see Section 3, page 109) to help us ride the wave of the emotion.
- How would you speak to someone you love who was in this scenario?
- How would you show them you love them while also being honest and enabling them to be accountable for their actions?
- Talk to someone you know and trust. Concealing shame keeps it going. Sharing it helps us to recognize the common human experience of shame after failure. Good friends can also help us to stay accountable for our mistakes because we trust them to be honest with us, while continuing to accept us throughout.
- What response to this situation most fits with the kind of person you want to be? How could you move forward from now in a way that you would be proud of and grateful for when you look back on this time?

Chapter summary

- How we think about stress effects how we perform under pressure.
- Seeing stress as an asset that will help you enables you to spend less energy trying to squash the feelings and instead focus on the demands you face.
- Keep performance affirmations or mantras focused on what to do, rather than what not to do.
- Adjust your focus to adjust your stress levels.
- Work on your relationship to failure and building shame resilience to help you deal with stress in high-pressure situations.

8

On a Meaningful Life

CHAPTER 32

The problem with 'I just want to be happy'

In therapy when we start to shine a light on the way forward and think about what we want, it's not uncommon to hear 'I just want to be happy.'

But the idea of happiness has been hijacked over the years by an elusive fairytale of constant pleasure and satisfaction with life. You don't have to look far on social media to come across a wave of posts telling you to 'be positive, stay happy, eliminate negativity from your life'.

We are given the impression that happiness is the norm and anything outside of that could be a mental health problem. We are also sold the idea that if we can achieve material wealth, happiness will arrive and stick around.

But humans are not built to be in a constant happy state. We are built to respond to the challenges of survival. Emotions are a reflection of our physical state, our actions, beliefs and what is going on around us. All of those things are constantly changing. Therefore, a normal state is one that constantly changes too. In his book *The Happiness Trap*, Russ Harris explains how emotions are like the weather. They are constantly moving and changing, sometimes predictably, sometimes suddenly and unexpectedly. Emotions are always a part of our experience. But, just like the weather, some moments are pleasant and others are hard to endure. At other times, there is nothing distinct enough for us to easily describe. When we recognize the nature of human experience in this way, it becomes clear that anything being sold to us with the promise of happily ever after cannot live up to its promise if happy means the absence of any of the less pleasant emotions. We can live a happy and fulfilling life and still experience the full range of emotions that comes along with being human. Buying into the idea that happiness means constant positivity can leave us believing we have failed when we feel down. We feel like we are getting something wrong, or we feel afraid that we may have a mental health problem. Thinking in that way then makes that dark cloudy day even darker. Sometimes we are not happy because we are human and life is difficult a lot of the time.

Things that bring us the most happiness in our lives bring much more than happy feelings. The best example is the people in our lives. Your family, who mean the world to you, may upset you the most when they get things wrong. Parents feel a profound sense of meaning in their roles and intense feelings of love and joy. But they also feel great pain and fear and shame at times too. So happy moments are just one flower in a very large bouquet. You can't have one without the other. Emotions come as a whole bunch.

Why meaning matters

Some people start therapy because they feel lost in life. They can't put their finger on a specific problem but they know they don't feel right. It's hard to get excited about anything or apply themselves to tasks with any real energy or enthusiasm. Without a clear and specific problem, they find it hard to problem-solve and work out which direction to take. It's not so much that they are struggling to achieve their goals. They are not sure which goals to set in the first place, and whether any of them feel worth it.

In many cases, this is associated with a disconnect from core values. Life has pulled them away from what matters most to them. Working to get real clarity on your values can do a number of things. It can give a guide on the direction you want to head in, an idea of the types of goals that will be most fulfilling and purposeful. It can help you to persevere through painful points in life and, crucially, to remind yourself that even when times are hard, you are on the right path.

What are values?

Values are not the same as goals. A goal is a concrete, finite thing that you can work towards. Once you achieve it, that is the end point. Then you have to look for the next goal. A goal might be passing an exam, ticking everything off your to do list, or running a personal best.

Values are not a set of actions that can be completed. Values are a set of ideas about *how* you want to live your life, the *kind* of person you want to be and the *principles* you want to stand for.

If life was one complete journey, then a value would be the path you choose to follow. The path never comes to an end. It is one possible way of making your journey, and living in line with your values is the conscious effort you make to always stay close to that path. The path is full of hurdles you have to jump over along the way. These are the goals you commit to when you choose this path. Some hurdles may be large and you're not even sure if you can get over them. But you give it your best shot because staying on this path is so important to you.

There are plenty of other paths with other hurdles and challenges. But choosing to stick to this one and tackle whatever comes along gives all those events and actions meaning and purpose. It is the intention of choosing this path that enables you to push through barriers that you might never have tried otherwise. So you may work hard to pass lots of exams in your lifetime, because one of your values might be lifelong learning and personal growth.

Values are the things you do, the attitude you do them with and why you choose to do them. They are not who you are and who you are not. They are not something you have or become or achieve or complete.

Sometimes we drift away from living in line with our values. That might be because life happens and we get pulled in different directions. Or it can be because we didn't have a clear sense of what our own values were. As we mature and develop throughout our life, our values may change too. We develop independence and move away from home, we learn from the people we encounter, learn more about the world, maybe have children, maybe don't. The list goes on. For all of these reasons, it is a valuable practice to engage in regular evaluation of what matters most. That way, we can make conscious decisions to redirect if we need to and ensure we stay close to that path so that life can feel meaningful.

When we don't have clarity on our values, we can set goals based on what we think we should be doing, others' expectations, or a guess that once we achieve that goal, we will finally be enough, we can finally relax and be happy with who we are. One major flaw with that is it puts rigid parameters around the conditions in which you can be content and happy. It also places life satisfaction and happiness all in the future (Clear, 2018).

I am not suggesting that you should never set yourself goals. But when you work towards something, it helps to be clear on why you are working towards it and to recognize that all the good in life is not waiting at the end point of goals, but in the process we go through along the way. Rather than hoping things are better in the future, what if life could be meaningful and purposeful now, by living in line with what matters most to you? You still get to strive towards change and achievement with all your strength, but you are not waiting for a meaningful life, you already have it.

Chapter summary

- We are often sold the idea that happiness is the norm and anything outside of that could be a mental health problem.
- Sometimes we are not happy because we are human and life is difficult.
- Things that make life worthwhile bring us more than just happy feelings. They bring us a mix of happiness, love, joy, fear, shame and hurt at times too.
- Getting clarity on our personal values can guide us on setting goals that will bring meaning and purpose.
- Keeping our values front and centre also helps us to persevere through painful points in life knowing we're on the right path.

CHAPTER 33

Working out what matters

There are some simple exercises you can do to get some clarity on your values as they are today. It's worth noting that values change over time depending on our stage in life and what we are facing. Not only do our values change but so do our actions and their alignment to those values. Life happens, and when we face change or struggle, we can be pulled in a new direction away from what matters. So it's helpful to make time for a values check-in every now and then to re-evaluate. It's a way of checking the compass and the map at the same time. Which way am I heading? Do I want to be going in this direction? If not, how can I adjust my direction to head back towards what matters most to me?

Figure 10: Values – circle the values that feel relevant and important to you.

```
ENTHUSIASM      HONESTY      FAITH      FAIRNESS

   KINDNESS      CARING      COMPASSIONATE

   STRENGTH      AMBITIOUS      DEPENDABLE

RELIABLE      PRESENT      FLEXIBILITY      CURIOSITY

     OPEN-MINDEDNESS      DARING      LOYALTY

   CREATIVITY      ADVENTUROUS      GRATITUDE

TRUSTWORTHY      UNDERSTANDING      SPIRITUALITY

   SUSTAINABILITY      SINCERITY      SELF-AWARE

   INDEPENDENCE      CONNECTION      ACCEPTING

     LOVING      DETERMINATION      PATIENCE

   PROFESSIONALISM      RESPECTFUL      BRAVERY
```

Figure 11: This chart gives a couple of examples of the distinction between values and goals that may be in line with those values and how that may translate into everyday actions.

VALUES	GOALS	DAY-TO-DAY BEHAVIOURS
Lifelong learning, curiosity, personal growth.	Educational courses.	Reading, studying, challenging self with exams or performances that push and expand those skills and promote learning.

Love and compassion for others.	To remember special dates for loved ones, to visit relatives at certain times.	Expressing love and compassion in small ways every day. Writing down dates of birthdays and anniversaries. Making time to spend with loved ones. Helping an elderly neighbour across the road.

 Toolkit: The values check-in

In the spare tools section at the back of the book, you'll find a blank grid that you can use to reflect on what you value most in each area of your life. The examples listed in these boxes are just starting points. You do not have to follow these exact ones. Feel free to change them for values and goals that most fit for you. In each box try to reflect on what values are most important to you in this area of your life. Here is a list of prompts that can help.

- What kind of person would you most like to be in this area of your life?
- What do you want to stand for?
- What do you want your efforts to represent?
- What contribution do you want to make?
- What qualities or attitude do you want to bring to this area of your life?

RELATIONSHIPS	HEALTH	CREATIVITY
PARENTING	SPIRITUALITY / FAITH	CONTRIBUTION
LEARNING AND DEVELOPMENT	PLAY / LEISURE	WORK

The key part of this exercise comes after you have listed your values in each box. In ACT therapy we ask individuals to rate how important each set of values are to them on a scale of 0–10.

On this scale, 10 would be of the highest importance and 0 not at all. We then ask people to rate how closely they feel they are living in line with those values on the same scale, 10 being highly in line with the value and 0 being not at all. We then spend time looking at the difference between the rating of importance and the rating of how closely they are living in line with it. If that difference is large, then it can indicate that you have been pulled away from living in line with what matters most to you. For example, if you identify that looking after your health and looking after your body is of the highest importance to you at 10 out of 10, but you rate how much you are living in line with that as just 2 out of 10 because you have been eating poorly and stopped exercising, then that gives you a prompt to look at making some positive changes in that area of your life.

All this does is give you an indication of areas of your life to turn towards. It is a great way to get a bird's-eye view of the sometimes competing priorities in our lives. It doesn't dictate what we should do or how to do it. It just offers us a map, an overview of things as they are now. From there, we get to choose what action we take to bring ourselves closer to the path that we want to be on.

Crucially, this exercise is not about all the problems we face and the pleasant or painful emotions we encounter each day. It is about the meaning we find in both the toughest and easiest days. It does not ask us to wait until everything is fine before we start living as the kind of person we want to be. It gets us thinking about how we can consciously choose to live by our values, whatever is going on around us.

Once you have identified a few of the most important aspects of your life and your values in that area, you can use this simple exercise to check in and see how closely you're living in line with your values at the moment. The exercise was originally designed by a Swedish ACT therapist named Tobias Lundgren. This is my own adaptation that I like to use myself.

The star shape has six measures, one on each point of the star. Label each measure with a domain of your life that is especially important to you. On the scale of 0-10, mark a cross that represents how much you are currently living in line with your values in this area. For example, you may feel that you haven't been prioritizing your health in the way that you'd like, so you give it a 5 out of 10. But in your relationships you might feel that you are living pretty closely in line with the kind of partner you want to be, so you give that one a 9 out of 10.

Once you have given them all a mark, you can draw lines to join the marks and see how your star takes shape. If the star is uneven, the shorter points are the areas that need your attention. You'll find blank values stars that you can fill in at the back of the book in the spare tools section.

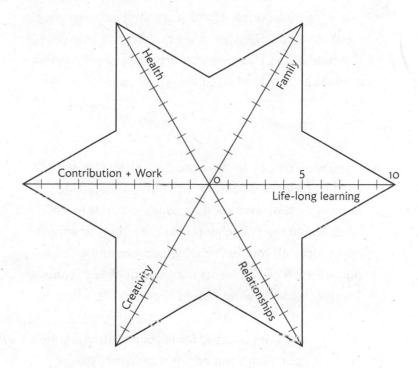

Figure 12: The values star.

It can be easy to feel confused about how much our values represent our own wishes and how much they are dictated by the expectations of others. This is an important issue to clarify. It is not to say that our sense of duty or commitment to our family or community is not important or that we should not choose it. But working out what values are truly your own and which ones feel imposed can reveal why certain aspects of our life might feel less fulfilling or disconnected.

 Try this: Another way to do a more regular values check-in is to incorporate it into journalling or some simple self-reflection. Here are a few questions to help prompt that reflection. I like to use these both personally and with others in therapy when exploring values.

Questions to explore:

1. If you were to look back on this next chapter of your life and feel proud and content with how you faced life's challenges, how would you be approaching daily life? What would the next chapter look like? In your answer try to focus on your own choices, actions and attitude, not other people or events that are out of your control. Try to consider how you would approach life, whatever happens.
2. What do you want to stand for in your relationship with yourself, your health and personal growth? What is important to you about these?

3. What kind of person do you want to be for the people in your life? How do you want to interact with them and contribute to their lives?

4. How do you want the people in your life to feel when you are around? What do you want to represent in your circle of friends and family?

5. If you only get to live once, what impact do you want to have while you are here?

6. If no one knew how you spent your time, would you still be doing this?

7. As you move forward through this day or this week, what is one value that you will try to bring to each choice and action? Examples here might be 'Today I choose to bring enthusiasm/courage/compassion/ curiosity to each experience, choice and action. I will do this by . . .'

Chapter summary

- There are some simple exercises you can do to get some clarity on your values as they are today.
- Values can change over time and how much we live in line with them can change. So it helps to do a regular values check-in.
- When we use our values to guide our goal-setting, it also helps to create our day-to-day purpose.
- The focus is not on what you want to happen for you but on the kind of person you want to be, the contribution you want to make and the attitude you want to face life with, no matter what happens.

CHAPTER 34

How to create a life with meaning

So what happens when you work out what is important and realize you are not living in line with those values? How do you begin to move in that direction? When we decide it's time for a change there can be a tendency to come up with a huge, radical new goal. For example, let's say you do your values check-in and decide you need to start looking after your health by exercising. In the next moment, you start coming up with new goals: maybe you'll run a marathon or improve your nutritional intake. But simply having goals is not going to ensure that your life changes and stays changed. What does that is the everyday details of your repeated behaviours that keep you moving forward in that direction.

Your goal might be to complete a marathon. But marathon

or no marathon, the life-changing part comes from the things you put in place to help you get out running every day, the running group you join to keep you going, the ways in which you gradually increase the distances you cover and the changes you make to your nutritional intake to help. Setting a goal can help to give you that initial push in the right direction. But it is important to remember that the end point of the goal, its completion, is actually a limitation. If you have reassessed your values and you want to move in a new direction based on what matters most to you in your life, then you are likely to want to continue in that direction. Many people run a marathon and hang up their running shoes shortly after.

The values check-ins are useful to do on a regular basis because the details of your values may change over time. But it also gives you the chance to focus on the intricate details of the everyday nitty-gritty of how you are living. We get to ask ourselves, 'What kind of person do I want to be *today*?' and 'What am I going to do *today* to step in that direction?' If your identity is going to be the kind of person who looks after their health every day, then that has potential to last long after the marathon has come and gone.

Working in this way is a two-pronged attack. By spending time thinking about and visualizing the person you intend to be and turning those ideas into concrete, sustainable actions, you can begin to change how meaningful those efforts will feel. Change is hard, so having a solid anchor in your clarity about why, and a permanent sense of identity, 'because this is who I am now', helps you to persist when that change inevitably meets resistance from your own mind or the people around you. Over time, once those new ways of thinking and behaving establish themselves, your beliefs about yourself can begin to shift too. So you genuinely become someone who prioritizes health and fitness, not by the initial goal of running a marathon, but by persisting with a new lifestyle. Exercise becomes something you do because you identify with it, not because there is a goal to reach. The initial marathon idea almost becomes irrelevant.

Overly focusing on outcome can lead us to quit more easily when we don't see results quickly enough or when we meet resistance and hurdles along the way. When you first decide to set a goal, you might get excited about it and a spark of motivation will strike. But motivation is like a flame on a match. It will burn itself out. It's an unsustainable source of fuel. But if you have a routine of small actions that are not too radical or dramatic to maintain, then your new sense of identity will help to sustain you.

Chapter summary

- When we decide it's time for a change there can be a tendency to come up with a huge radical new goal.
- Simply having a goal is not enough to ensure your life changes and stays changed.
- Spending time thinking about and visualizing the person you intend to be, and turning those ideas into concrete, sustainable actions, can change how meaningful those efforts feel.
- Linking your intentions to your sense of identity allows the new behaviours to continue way beyond the initial goal.

CHAPTER 35

Relationships

We cannot talk about a life of meaning without talking about relationships. Our relationships are what make us human. When it comes to living a happy life, relationships trump money, fame, social class, genes and all the things we are told to strive for first and foremost. Our relationships and how happy we are in them are not separate from our overall health. They are at the core of the equation. Healthy relationships protect both our physical and mental health over the course of our lives (Waldinger, 2015). This does not just mean life partners and marriage, but all our relationships. Those with friends, family, children and our community. This comes up in both the scientific data on various health measures and biological indicators, but also in the narratives of real people. Among the top five regrets of the dying is 'I wish I had stayed in touch with my friends' (Ware, 2012).

But for something so profoundly defining about who we are and how we live our lives, and something so powerful in its impact on how long and how happily we live, we are all left guessing as to what we should be doing to make our relationships healthy. Nobody hands us a manual.

We start connecting with others and learning from those experiences from the moment we are born. We build up templates for relationships from the very first relationships we have, with parents, siblings, extended family members and peers. We have to learn these lessons at our most vulnerable age, when we cannot choose our relationships but are wholly dependent on them for our survival.

Those patterns of behaviour that we learn to use early in life to manage our relationships can sometimes prove much less helpful to us in our adult relationships.

But, given that relationships are so crucial to a long and happy life, how can we start (even as adults) to work out how to improve them?

Insights from both individual therapy and couples therapy can help us with this. Cognitive Analytic Therapy (CAT) is a therapy that acknowledges the relationship patterns that we develop early in life and how those play out in our adult relationships. For those able to access CAT, it can be a revealing process to map out the roles you tend to find yourself playing in relationships and the cycles you feel stuck in.

But for those of us who cannot access a therapy like CAT, what can we do to better understand our relationships and work on improving them?

First, it is important to point out some of the myths we are led to believe by popular culture that can leave us feeling that we are getting things seriously wrong. Most of these myths stand for both intimate relationships and others with friends and family.

Relationship myths

- **Love shouldn't be hard.** The idea that if someone is right for you then the two of you drift off into the sunset and everything should be fine all of the time has no bearing on real life and leaves most people feeling dissatisfied with their relationships. A long-lasting relationship is not a gentle boat ride that drifts downstream. You have to pick up the oars and make values-based choices and actions about where you want to go with it. Then you have to put the work in. Those actions have to be repeated consistently. If you spend more time drifting than purposefully choosing and working on it, things can get off course.
- **Be as one.** In a relationship or friendship, it is perfectly OK to disagree. You don't have to be on the same page about everything all of the time. You are two different people, each with your own sensitivities, background experiences, needs and coping mechanisms. If you truly open up and connect with another person, you will undoubtedly discover parts of them that you need to tolerate and accept in order to nurture the relationship over a lifetime.
- **Be together always.** Whether it's a friendship or intimate relationship, it is OK to enjoy spending time

apart. You do not have to become two parts of the same person. You are two separate and unique individuals and to nurture the aspects of yourself that make you different does not need to threaten the relationship. This relationship myth compounds our fears of abandonment and prevents many people from allowing their partners or themselves to develop and grow as individuals within a relationship. When we feel secure in a relationship, we can feel more free to be separate people and not feel threatened by the other aspects of our partner's life.

- **Happily ever after.** From fairytales to Hollywood movies, the story always ends as the relationship begins, as if the journey is only in finding the perfect person and after that is endless happiness. A relationship is a journey that will naturally meet with many twists, turns and bumps in the road. The strongest of relationships will have down days, periods of disconnection, disagreement. There will be times when one or both partners face failure or huge loss, or illness and pain. There will be times when you have mixed emotions or you feel less passionate than before. There will be times when one or both partners feel confused about what the other person wants or needs. There will be times when we get it wrong and cause the other person pain. If we buy into the myth of happily ever after, we make ourselves vulnerable to assuming that this relationship just wasn't meant to be and end it without realizing that all relationships hit bumps in the road. When they knock you over, getting back up and coming back together is possible.

- **Relationship success means staying together at all costs.** Relationships have a potent effect on our health and happiness, but merely having a relationship is not enough. If relationships are to have a positive impact on our lives, then this means working to improve the quality of those connections and making careful, intentional choices about them too. While we can take full responsibility for ourselves, we cannot force change upon another. It is OK to end a relationship that causes harm to your physical or mental wellbeing. I have included in the resources section information on services that provide support to anyone who feels unsafe in a relationship.

How to get better at relationships

When we look after ourselves we are looking after our relationships, and when we work on our relationships we are looking after ourselves. So all the tools in this book that focus on taking care of the self will help you to be the person you want to be in your relationships.

Getting better at relationships does not mean learning how to get the other person to do or be what you want them to be. In couples therapy, you can work on your relationship together. But you can also work on your relationships by understanding your own individual needs and patterns and the cycles you tend to get stuck in. When you build a better understanding of yourself and practise new ways of communicating and connecting with the people in your life (including you), you can make real shifts in the quality of your relationships. That understanding of who you want to be, how you want to be there for the people in your life and how to hold boundaries and nurture yourself within those relationships can act as a compass. So when we feel lost and confused in the complexities of relationship ups and downs, we don't have to look to others for our sense of direction. We come back to the self. We step back from the painting and we see how our current choices fit with the bigger picture we are trying to create.

Attachment

Our attachment styles are formed early in life. They do not start off as choice. The brain is wired to attach to a caregiver to keep us safe. It enables every child to seek closeness with a parent, go to that parent for safety and comfort when needed, and use that relationship to create a secure base. When a child has that secure base, the child then feels safe to explore the world and form new relationships using what they have learned. But when life happens and parents are not able to give the consistent connection and security that children need to develop that secure attachment, then we can carry those insecure internal processes into our adult relationships (Siegel & Hartzell, 2004).

They impact on the way we relate to others as adults because they are the template we have formed for our concept of what to expect from and how to behave in relationships. Having a particular attachment style is not a life sentence that sets in stone how you are destined to relate to others. But it can be a helpful way to understand some of the cycles we feel stuck in as adults. Our brains are adaptable, so understanding those patterns and making conscious choices to repeatedly do something different, eventually can become our new norm.

Anxious attachment

An anxious attachment style may present itself as the need for frequent reassurance that you are loved and that the other person is not about to abandon you. Those with an anxious attachment style may have grown up in an environment in which they did not feel safe that their caregivers would return, or one where they did not have access to consistent affection or responsiveness and availability was inconsistent.

Anxious attachments can show up in people-pleasing behaviours; struggles in expressing personal needs or avoidance of confrontation and conflict, a focus on meeting the needs of the partner to the detriment of one's own personal needs.

The constant focus on preventing abandonment can become a self-fulfilling prophecy, as the relentless need for reassurance can feel controlling to those who may have an avoidant attachment style, and may lead to conflict. The anxiously attached partner may build up resentment towards a partner who does not consistently provide that reassurance but may also feel unable to fully express their needs for fear of conflict.

The answer in these situations is neither a constant flow of reassurance, nor dismissing the needs and hoping they go away. Instead, the anxiously attached partner can practise creating a sense of safeness independently from their partner by building their sense of self and learning to soothe themselves. The partner can help this by providing more consistent connection without waiting for the other person to ask for it. So these are things we can work on both individually and as a couple.

Avoidant attachment

This attachment style can appear almost opposite to an anxious attachment style. Closeness and intimacy can feel threatening and unsafe, despite still having that human need for connection. Self-reliance feels safe, giving just enough of oneself to keep a relationship but experiencing discomfort, vulnerability and fear in those connections and struggling with urges to shut down emotionally and avoid intimacy or confrontation.

These behaviours can often be mistaken for a lack of love or care. But they can be understood as something that once made good sense for the individual. Those with avoidant style attachments may have experienced a childhood in which parents weren't available either physically or emotionally to connect and respond to their needs. Dependence may have led to rejection or caregivers may have been unresponsive.

There is a misconception that those with an avoidant attachment style do not want or need connection. They are just as human as anyone else, but miss out on deep connection in the struggle to let down the protective guards that were put up much earlier in life to protect them. While the anxiously attached person must work on tolerating the vulnerability of self-reliance, the avoidantly attached must build up tolerance of the vulnerability involved in opening up to close connection. A partner can help with that by developing an understanding of why intimacy feels unsafe or uncomfortable and working alongside them to gradually nurture closeness.

Secure attachment

When parents are able to reliably respond to their child's emotional and physical needs, that child is able to learn over time that what she feels can be communicated, and responded to. She feels safe to express needs and learns that she is able to go into the world to get those needs met. This does not mean parenting was perfect, but it was reliable enough to generate that secure base and was repaired when mistakes were made so that trust continued.

The securely attached child is not constantly happy, with every need anticipated before a cry. They will feel safe enough to show their distress when a parent leaves, but will rekindle that connection when reunited. As they continue into adulthood they will enjoy closeness, feel able to express their needs and feelings, while maintaining the capacity for some independence.

A secure attachment is a solid foundation for managing healthy relationships as an adult, but is not a guarantee of ideal relationship choices or behaviours. Those with a secure attachment style navigating a relationship with someone who has a different attachment style can improve their relationship by working hard to understand and show compassion to the partner who had a different experience in their early years.

Disorganized attachment

If parents are unable to provide care and emotional support that is reliable and consistent, or if that relationship is abusive, then these kinds of interactions may create a disorganized attachment. As a child, that might be seen as avoidant or resistant responses to caregivers because the mix of experiences is confusing and disorienting. The person they need for safety can also be frightening and dangerous. Later on, in adulthood, this attachment style might show up as difficulties in dealing with emotion and a vulnerability towards dissociation in response to stress, intense fear of abandonment and difficulties in relationships.

As with the other attachment styles, change is possible with support. There may be work to do on building up the ability to manage both the vulnerability of intimacy and connection, as well as the fears of separation.

While the experiences we had in early life can be highly potent in how we express ourselves in adult relationships, they do not have to be a life sentence. Learning about ourselves and those we are closest to is the work of relationships. Recognizing and building an understanding around our own patterns of relating, as well as those of the people we may be in a relationship with, is a big step towards improving our relationships. It increases our chances of being able to step back from personalizing the behaviour of others and making conscious choices that can help you both build a close and trusting connection that enhances both of your lives.

So, how can we put this into action? What can we do to start improving our relationships today? As with most things, there are no quick fixes. Creating anything that is long-lasting is not about one big grand gesture that will make everything OK. It is about making those seemingly small everyday choices conscious and intentional. It is about reliably and consistently trying to navigate towards your values. The way to ensure that your everyday actions are led by intention, rather than reactive, is to step back every now and then and reflect on how you want the picture to look.

Relationships researcher John Gottman (Gottman & Silver, 1999) suggests that for both men and women the overriding factor that determines how satisfied they feel in their relationship (by 70 per cent) is the quality of their friendship. So actively focusing on how to develop friendship and working on what it takes to be a better friend is a good idea.

When we work on building the quality of friendship, we might do that by regularly enjoying each other's company and working hard to maintain mutual compassion and respect for each other, getting to know each other in the finest detail, finding ways to express appreciation and care in everyday life. The more of life we can fill with closeness and experiences that strengthen friendship, the more protection the relationship has against the inevitable hurdles that come in the form of disagreements, stressful life events and losses. It is much easier to ride the waves of life's ups and downs if we are well-practised at pulling together and have built up deep respect and gratitude for each other.

Connection

In this book I have talked a lot about the dangers of numbing emotions and avoidance. Relationships, whether they are intimate or friendships or family members, are all inherently woven with emotion. When humans interact, we make each other feel things. A few words from a loved one can put our head in the clouds or bring us to our knees. It makes sense that when the emotions are heightened, we pull back and withdraw. When we do, there is disconnection between the two individuals. And yet, any couples therapist and the research literature will tell you that turning towards each other is the foundation for building a deep and trusting connection (Gottman & Silver, 1999).

Disconnection from our selves, our emotions and our loved ones has negative outcomes for relationships and our mental health (Hari, 2018). Yet we are surrounded by things that tempt us with escape from the vulnerable moments. We numb out with endless social media scrolling or we throw ourselves into work and make ourselves too busy to stop. Or we turn away from our connection in order to obsess over making ourselves better along the lines of what the outside world tells us will do that. We focus on trying to look something closer to perfect, or reach somewhere closer to wealthy. None of which is what our connection truly needs to make it work.

So what does work? Here is what the experts say about ways in which we can build a meaningful and long-lasting connection.

- **Self-awareness:** Relationships are difficult because we don't always have access to what someone else is needing or thinking or feeling. But we can access that in ourselves. The most powerful place to start on improving your relationships is with you. Not in a self-blame, self-attack crusade. But with curiosity and compassion. Understanding the cycles that we seem to get stuck in and what may have made us vulnerable to that. This paves the way to working out how to break those cycles. We cannot always guarantee that the other person in a relationship will be inclined to self-reflect in this way. But when we start to make changes to our behaviour, it can invite the other person to respond differently too. This doesn't mean changing and hoping they change too. It means focusing on who you want to be in the relationship, how you want to behave and what you want to bring to the connection, as well as where your boundaries are and why.

- **Emotional responsiveness:** The big emotions we feel when a relationship hits trouble are not irrational. Safe emotional connection is a top priority for our brain, whose job is to help us survive. When we shout, scream, cry, withdraw and stay silent, we are all asking the same thing, just in a different way. The questions are 'Are you there for me? Do I matter enough for you to stay? When I need you most, what are you going to do?' The attachment styles we discussed earlier are the different ways in which we learned to ask those questions. When we sense that we have lost our connection, our brain sets the fight-or-flight alarm bells ringing and we start doing whatever we can to feel safe again. For some that is aggression, for others it is backing off and hiding away, or bringing down the emotional shutters and not letting on that you care at all. Once we get into a cycle of attacking and pulling away, it feels almost impossible to come back together, despite the fact that the disconnection has caused the distress. In her book *Hold Me Tight*, Sue Johnson, professor of clinical psychology and expert on Emotionally Focused Couple Therapy, suggests that if we don't reconnect, we continue feeling isolated and distanced. The only way back is to move emotionally close and reassure the other. She points out that while one partner might blame or attack in a frantic attempt to seek an emotional response, the other person is likely to receive the message that they are failing and freeze or retreat further away. To remedy this, we can practise tuning into our partner's bids for connection and attachment needs. This is easier said than done when overwhelmed with emotion, so it inevitably involves also working on self-soothing and managing our own distress too. Secondly, it includes responding to those attachment cues with sensitivity, kindness and compassion, letting the other person know that they matter to you. It is crucial that while we do this, we remain engaged, close and attentive rather than pulling away (Johnson, 2008).

- **Respectful complaining:** Most people have an idea of what type of feedback helps them to take a message on board and learn, and what type of feedback puts them in a shame spiral. When we play the blame game, nobody wins. Building a healthy relationship is not about letting go of your own needs to please another, but it does demand that we use the compassion and care that we would want to receive when we tackle frustrations and problems.

  Healthy relationships are not free of conflict. They require work on carefully repairing ruptures in the connection. Whatever the details of each particular conflict, each individual continues to have the same basic need for feeling love and belonging, a sense of acceptance for who they are as separate from mistakes or unhelpful behaviour patterns. One of the fundamental aspects of therapy that creates a solid foundation for the ability to self-reflect and work on change is the creation of a relationship in which there is acceptance, non-judgement and unconditional positive regard. When we sense that we are under attack or being abandoned, when we feel shamed or unappreciated, we are not in a position to think clearly about what would be the best way forward. We are in survival mode. When tackling a difficult conversation, careful thought and preparation in how you will approach the conversation is more likely to go smoothly than allowing frustration to lead the way with a stream of criticism and contempt. Focusing use of language on concrete behaviours rather than global attacks on personality helps to keep everyone cool. Staying clear on exactly what it is you feel and need helps to keep the guessing games out of it, while maintaining the appreciation and respect that you would want to be treated with if the roles were reversed is a good place to start. Of course, none of this is easy, especially when emotions are high, so it demands that we keep returning to our own personal values around the kind of partner we want to be.

- **Repairs:** When it comes to repairing, our priority is reconnection. This inevitably involves acknowledgement of our role in what happened, compromise and adjustments from both individuals. Reconnection also demands the ingredients that created the connection in the first place: acceptance, compassion, love and gratitude for each other. Accessing that is almost impossible when emotions are high, so it doesn't have to happen immediately. It is OK to minimize damage by taking a short time to step back and calm down, before re-approaching in a more skilful way. All of this sounds idealistic, and real life doesn't always happen that way. Old habits are pretty hard to crack. But there is no use in being a relationship perfectionist. Sometimes we will get it wrong. The key is in persistence and the commitment to take a step back, re-evaluate and do our best to repair when things go wrong. Anything repeated enough times will become second nature.

- **Turning towards gratitude:** In previous chapters I have talked about the value of redirecting the focus of attention towards gratitude. In the hustle and bustle of daily life, it is easy to slip into a pattern of paying most attention to a partner when we need them to step up and change something, or when they are frustrating the hell out of us. Making a conscious decision to focus on the things we admire and appreciate about them is a relatively simple task that can shift not only your emotional state but how you then choose to behave towards them.

- **Shared meaning and values:** If we choose to spend our life with another person, then our values check-in, our stepping back to see our bigger picture, cannot be ours alone. Finding where our personal values fit together and overlap with our partner's and having respect for where they differ is key for a relationship that can withstand life's challenges. This might start with the relationship, how you both want to care and be cared for, communicate and be communicated to, support and be supported. It might also expand to consider both personal goals and shared dreams for your life together. There may be aspects of your relationship and family life that are sacred to you both, while there will be others where one partner upholds a value because she knows how important it is to the other. For example, going along to family gatherings with relatives that may not be your favourite because you know how much it means to your partner that you are there to support him. As I described in earlier chapters, gaining clarity on what matters most to you acts as a compass and a guide when we are unsure how to move forward. When we are in a relationship, taking time to understand what matters most to our partner can help us to deepen our connection and create a relationship within which both partners can grow and flourish.

 Toolkit: Getting clear on the kind of partner you want to be

The following journal prompts can help to explore your shared values as a couple. You can use these to reflect on any of the relationships in your life. As we can't force change upon others, the focus is on understanding and identifying what we can do as individuals.

- Which of the attachment styles listed in this chapter resonate for you?
- How does that appear in your relationships?
- How can you express compassion for the unintended consequences of those past experiences, while also taking responsibility for your future?
- What aspects of your partner and the relationship do you feel appreciative and grateful for?
- What kind of partner do you want to be in the relationship?
- What small changes could help you move in that direction?

Chapter summary

- When it comes to a happy life, relationships beat money, fame, social class and all the things we are told to put our effort into.
- Our relationships and how happy we feel in them are not separate from our overall health. They are at the core of the equation.
- Working on the self helps your relationships, and working on your relationships helps the self.
- Attachment styles early in life can often be reflected in our adult relationships.

CHAPTER 36

When to
seek help

Dear Dr Julie,

I saw your videos. I was inspired to start therapy. So far it's going really well and things are starting to improve for me.

Thank you.

If anyone is wondering why it's important to talk about mental health, here is one reason. In my first year of offering mental health education online, I lost count of the number of messages I received that sounded like the one above. In each one the words are different, the stories unique. But the message is the same. And it's not just me. There are people talking about mental health and therapy all over the internet. On an individual level, this is what it can do.

When your mental health fluctuates, it can be even more

difficult to make decisions and take action. So seeking the help you need becomes harder to do. And there is no set of rules that tells you when to see a professional.

When to seek professional help with your mental health is a question I am asked frequently. The short answer to this question is any time you are concerned about your mental health.

There are huge barriers for many people in the world in accessing professional mental health support. From cultural taboos and expensive services to availability and resources, there are very real hurdles that prevent huge numbers of people accessing services that could be helpful to them. Overcoming each of those hurdles is a massive challenge to be faced by society. On an individual level, if you are fortunate enough to have the chance to access services and you feel in any way concerned about your mental health, then taking that step can be life-changing. Simply visiting a professional and starting the conversation allows you to explore your options.

Something that I often hear from individuals when they come along to talk about the idea of therapy is that they don't feel they deserve it. Other people must have it worse. So they wait until breaking point before taking that step. By that time, the hill to climb has become a mountain. Waiting until you are on your deathbed before seeking help is never a good strategy for maintaining your health, both physically and mentally. The truth is there will always be someone who has it worse. But if you have the chance to use professional help along the way, your mental health may thank you for it and life could change beyond your current comprehension. Believe me. I have seen it happen. I have seen people pull themselves from the depths of despair, step back from the cliff edge and start the work of turning their life around. It happens, and it could happen to you. Not in a day or a week. But in many days and many weeks of commitment to your health and to the life you want to build.

When there is no way to access professional help, we need each other more than ever. The internet has made lots of educational resources more available and started a global conversation about mental health. People who once felt alone in their struggles are beginning to understand that fluctuations in mental health, just like physical health, are a normal part of being human. Stories of recovery, healing and growth are being told. Seeds of hope are being sown. The message is starting to be heard that our mental health is not entirely out of our hands. We are not at the mercy of emotional states that strike us down. There are things we can learn, changes we can make to take responsibility for our health. That involves learning all that you can from whatever is available to you and working hard to try things out, make mistakes, try again, learn a bit more and keep going.

In an ideal world, all the therapies that work would be available to everyone who needs them, when they need them. But we don't have that ideal world. So if professional services are not available, take every opportunity you can to learn and to share with people you trust. Human connection and education can help us to make big changes in our mental health.

Chapter summary

- The best time to seek support for your mental health is any time you become concerned about it.
- If you are not sure how much help you need, a professional can help you decide.
- In an ideal world, therapeutic services would be available to everyone. But we don't have that ideal world.
- If services are not accessible for you, take every opportunity to learn all you can about recovery and to use the support of trusted loved ones.

References

Section 1: On Dark Places

Beck, A. T., Rush, A. J., Shaw, B. F, & Emery, G. (1979), *Cognitive Therapy of Depression*, New York: Wiley.

Breznitz, S., & Hemingway, C. (2012), *Maximum Brainpower: Challenging the Brain for Health and Wisdom*, New York: Ballantine Books.

Brown, S., Martinez, M. J., & Parsons, L. M. (2004), 'Passive music listening spontaneously engages limbic and paralimbic systems', *Neuroreport, 15* (13), 2033–7.

Clark, I., & Nicholls, H. (2017), *Third Wave CBT Integration for individuals and teams: Comprehend, cope and connect*, London: Routledge.

Colcombe, S., & Kramer, A. F. (2003), 'Fitness effects on the cognitive function of older adults. A meta-analytic study', *Psychological Science, 14* (2), 125–30.

Cregg, D. R., & Cheavens, J. S., 'Gratitude Interventions: Effective Self-help? A Meta-analysis of the Impact on Symptoms of Depression and Anxiety', *Journal of Happiness Studios* (2020), https.//doi.org/10 1007/s10902-020-00236-6

DiSalvo, D. (2013), *Brain Changer: How Harnessing Your Brain's Power to Adapt Can Change Your Life*, Dallas: BenBella Books.

References

Feldman Barrett, L. (2017), *How Emotions Are Made. The Secret Life of The Brain*, London: Pan Macmillan.

Gilbert, P. (1997), *Overcoming Depression: A self-help guide to using Cognitive Behavioural Techniques*, London: Robinson.

Greenberger, D., & Padesky, C. A. (2016), *Mind over Mood, 2nd Edition*, New York: Guilford Press.

Inagaki, Tristen, K., & Eisenberger, Naomi I. (2012), 'Neural Correlates of Giving Support to a Loved One', *Psychosomatic Medicine, 74* (1), 3–7.

Jacka, F. N. (2019), *Brain Changer*, London: Yellow Kite.

Jacka, F. N., et al. (2017), 'A randomized controlled trial of dietary improvement for adults with major depression (the 'SMILES' trial)', *BMC Medicine, 15* (1), 23.

Josefsson, T., Lindwall, M., & Archer, T. (2013), 'Physical Exercise Intervention in Depressive Disorders: Meta Analysis and Systemic Review', *Medicine and Science in Sports, 24* (2), 259–72.

Joseph, N. T., Myers, H. F., et al. (2011), 'Support and undermining in interpersonal relationships are associated with symptom improvement in a trial of antidepressant medication', *Psychiatry, 74* (3), 240–54.

Kim, W., Lim, S. K., Chung, E. J., & Woo, J. M. (2009), 'The Effect of Cognitive Behavior Therapy-Based Psychotherapy Applied in a Forest Environment on Physiological Changes and Remission of Major Depressive Disorder', *Psychiatry Investigation, 6* (4), 245–54.

McGonigal, K. (2019), *The Joy of Movement*, Canada: Avery.

Mura, G., Moro, M. F., Patten, S. B., & Carta, M. G. (2014), 'Exercise as an Add-On Strategy for the Treatment of Major Depressive

Disorder: A Systematic Review', *CNS Spectrums, 19* (6), 496–508.

Nakahara, H., Furuya, S., et al. (2009), 'Emotion-related changes in heart rate and its variability during performance and perception of music', *Annals of the New York Academy of Sciences, 1169*, 359–62.

Olsen, C. M. (2011), 'Natural Rewards, Neuroplasticity, and Non-Drug Addictions', *Neuropharmacology, 61* (7), 1109–22.

Petruzzello, S. J., Landers, D. M., et al. (1991), 'A meta-analysis on the anxiety-reducing effects of acute and chronic exercise. Outcomes and mechanisms', *Sports Medicine, 11* (3), 143–82.

Raichlen, D. A., Foster, A. D., Seillier, A., Giuffrida, A., & Gerdeman, G. L. (2013), 'Exercise-Induced Endocannabinoid Signaling Is Modulated by Intensity', *European Journal of Applied Physiology, 113* (4), 869–75.

Sanchez-Villegas, A., et al. (2013), 'Mediterranean dietary pattern and depression: the PREDIMED randomized trial', *BMC Medicine, 11*, 208.

Schuch, F. B., Vancampfort, D., Richards, J., et al. (2016), 'Exercise as a treatment for depression: A Meta-Analysis Adjusting for Publication Bias', *Journal of Psychiatric Research, 77*, 24–51.

Singh, N. A., Clements, K. M., & Fiatrone, M. A. (1997), 'A Randomized Controlled Trial of the Effect of Exercise on Sleep', *Sleep, 20* (2), 95–101.

Tops, M., Riese, H., et al. (2008), 'Rejection sensitivity relates to hypocortisolism and depressed mood state in young women', *Psychoneuroendocrinology, 33* (5), 551–9.

References

Waldinger, R., & Schulz, M. S. (2010), 'What's Love Got to Do With It?: Social Functioning, Perceived Health, and Daily Happiness in Married Octogenarians', *Psychology and Aging, 25* (2), 422–31.

Wang, J., Mann, F., Lloyd-Evans, B., et al. (2018), 'Associations between loneliness and perceived social support and outcomes of mental health problems: a systematic review', *BMC Psychiatry, 18*, 156.

Watkins, E. R., & Roberts, H. (2020), 'Reflecting on rumination: Consequences, causes, mechanisms and treatment of rumination', *Behaviour, Research and Therapy, 127*.

Section 2: On Motivation

Barton, J., & Pretty., J. (2010), 'What is the Best Dose of Nature and Green Exercise for Improving Mental Health? A Multi-Study Analysis', *Environmental Science & Technology, 44*, 3947–55.

Crede, M., Tynan, M., & Harms, P. (2017), 'Much ado about grit: A meta-analytic synthesis of the grit literature', *Journal of Personality and Social Psychology, 113* (3), 492–511.

Duckworth, A. L., Peterson, C., Matthews, M. D., & Kelly, D. R. (2007), 'Grit: Perseverance and passion for long-term goals', *Journal of Personality and Social Psychology, 92* (6), 1087–1101.

Duhigg, C. (2012), *The Power of Habit: Why we do what we do and how to change*, London: Random House Books.

Gilbert, P., McEwan, K., Matos, M., & Rivis, A. (2010), 'Fears of Compassion: Development of Three Self-Report Measures', *Psychology and Psychotherapy, 84* (3), 239–55.

Huberman, A. (2021), Professor Andrew Huberman describes the biological signature of short-term internal rewards on his podcast and YouTube channel, The Huberman Lab.

Lieberman, D. Z., & Long, M. (2019), *The Molecule of More*, BenBella Books: Dallas.

Linehan, M. (1993), *Cognitive-Behavioral Treatment of Borderline Personality Disorder*, Guildford Press: London.

McGonigal, K. (2012), *The Willpower Instinct*, Avery: London.

Oaten, M., & Cheng, K. (2006), 'Longitudinal Gains in Self-Regulation from Regular Physical Exercise, *British Journal of Health Psychology, 11*, 717–33.

Peters, J., & Buchel, C. (2010), 'Episodic Future Thinking Reduces Reward Delay Discounting Through an Enhancement of Prefrontal-Mediotemporal Interactions', *Neuron, 66*, 138–48.

Rensburg, J. V., Taylor, K. A., & Hodgson, T. (2009), 'The Effects of Acute Exercise on Attentional Bias Towards Smoking-Related Stimuli During Temporary Abstinence from Smoking', *Addiction, 104*, 1910–17.

Wohl, M. J. A., Psychyl, T. A., & Bennett, S. H. (2010), 'I Forgive Myself, Now I Can Study: How Self-forgiveness for Procrastinating Can Reduce Future Procrastination', *Personality and Individual Differences, 48*, 803–8.

Section 3: On Emotional Pain

Feldman Barrett, L. (2017), *How Emotions Are Made. The Secret Life of The Brain*, London: Pan Macmillan.

Inagaki, Tristen, K., & Eisenberger, Naomi I. (2012), 'Neural Correlates of Giving Support to a Loved One', *Psychosomatic Medicine, 74* (1), 3–7.

References

Kashdan, T. B., Feldman Barrett, L., & McKnight, P. E. (2015), 'Unpacking Emotion Differentiation: Transforming Unpleasant Experience By Perceiving Distinctions in Negativity', *Current Directions In Psychological Science, 24* (1), 10–16.

Linehan, M. (1993), *Cognitive-Behavioral Treatment of Borderline Personality Disorder*, London: Guildford Press.

Starr, L. R., Hershenberg, R., Shaw, Z. A., Li, Y. I., & Santee, A. C. (2020), 'The perils of murky emotions: Emotion differentiation moderates the prospective relationship between naturalistic stress exposure and adolescent depression', *Emotion, 20* (6), 927–38. https://doi.org/10.1037/emo0000630

Willcox, G. (1982), 'The Feeling Wheel', *Transactional Analysis Journal, 12* (4), 274–6.

Section 4: On Grief

Bushman, B. J. (2002), 'Does Venting Anger Feed or Extinguish the Flame? Catharsis, Rumination, Distraction, Anger, and Aggressive Responding', *Personality and Social Psychology Bulletin, 28* (6), 724–31.

Kubler-Ross, E. (1969), *On Death and Dying*, New York: Collier Books.

Rando, T. A. (1993), *Treatment of Complicated Mourning*, USA: Research Press.

Samuel, J. (2017), *Grief Works. Stories of Life, Death and Surviving*, London: Penguin Life.

Stroebe, M. S., & Schut, H. A. (1999), 'The Dual Process Model of Coping with Bereavement: Rationale and Description', *Death Studies, 23* (3), 197–224.

Worden, J. W., & Winokuer, H. R. (2011), 'A task-based approach for counseling the bereaved'. In R. A. Neimeyer, D. L. Harris, H. R. Winokuer & G. F. Thornton (eds.), *Series in Death, Dying and Bereavement. Grief and Bereavement in Contemporary Society: Bridging Research and Practice*, Abingdon: Routledge/Taylor & Francis Group.

Zisook, S., & Lyons, L. (1990), 'Bereavement and Unresolved Grief in Psychiatric Outpatients', *Journal of Death and Dying, 20* (4), 307–22.

Section 5: On Self-doubt

Baumeister, R. F., Campbell, J. D., Krueger, J. I., & Vohs, K. D. (2003), 'Does High Self-esteem Cause Better Performance, Interpersonal Success, Happiness, or Healthier Lifestyles?', *Psychological Science in the Public Interest, 4* (1), 1–44.

Clark, D. M., & Wells, A. (1995), 'A cognitive model of social phobia'. In R. R. G. Heimberg, M. Liebowitz, D. A. Hope & S. Scheier (eds.), *Social Phobia: Diagnosis, Assessment and Treatment*, New York: Guilford Press.

Cooley, Charles H. (1902), *Human Nature and the Social Order*, New York: Scribner's, 183–4 for first use of the term 'looking glass self'.

Gilovich, T., Savitsky, K., & Medvec, V. H. (2000), 'The spotlight effect in social judgment: An egocentric bias in estimates of the salience of one's own actions and appearance', *Journal of Personality and Social Psychology, 78* (2), 211–22.

Gruenewald, T. L., Kemeny, M. E., Aziz, N., & Fahey, J. L. (2004), 'Acute threat to the social self: Shame, social self-esteem, and cortisol activity', *Psychosomatic Medicine, 66*, 915–24.

Harris, R. (2010), *The Confidence Gap: From Fear to Freedom*, London: Hachette.

Inagaki, T. K., & Eisenberger, N. I. (2012), 'Neural Correlates of Giving Support to a Loved One', *Psychosomatic Medicine, 74*, 3–7.

Irons, C., & Beaumont, E. (2017), *The Compassionate Mind Workbook*, London: Robinson.

Lewis, M., & Ramsay, D. S. (2002), 'Cortisol response to embarrassment and shame', *Child Development, 73* (4), 1034–45.

Luckner, R. S., & Nadler, R. S. (1991), *Processing the Adventure Experience: Theory and Practice*, Dubuque: Kendall Hunt.

Neff, K. D., Hseih, Y., & Dejitthirat, K. (2005), 'Self-compassion, achievement goals, and coping with academic failure', *Self and Identity, 4*, 263–87.

Wood, J. V., Perunovic. W. Q., & Lee, J. W. (2009), 'Positive self-statements: Power for some, peril for others', *Psychological Science, 20* (7), 860–66.

Section 6: On Fear

Frankl, V. E. (1984), *Man's Search for Meaning: An Introduction to Logotherapy*, New York: Simon & Schuster.

Gesser, G., Wong, P. T. P., & Reker, G. T. (1988), 'Death attitudes across the life span. The development and validation of the Death Attitude Profile (DAP)', *Omega, 2*, 113–28.

Hayes, S. C. (2005), *Get Out of Your Mind and Into Your Life: The New Acceptance and Commitment Therapy*, Oakland, CA: New Harbinger.

Iverach, L., Menzies, R. G., & Menzies, R. E. (2014), 'Death anxiety and its role in psychopathology: Reviewing the status of a

transdiagnostic construct', *Clinical Psychology Review, 34,*
580–93.

Neimeyer, R. A. (2005), 'Grief, loss, and the quest for meaning',
Bereavement Care, 24 (2), 27–30.

Yalom. I. D. (2008), *Staring at the Sun: Being at peace with your
own mortality*, London: Piatkus.

Section 7: On Stress

Abelson, J. I., Erickson, T. M., Mayer, S. E., Crocker, J., Briggs, H.,
Lopez-Duran, N. L., & Liberzon, I. (2014), 'Brief Cognitive
Intervention Can Modulate Neuroendocrine Stress Responses to
the Trier Social Stress Test: Buffering Effects of Compassionate
Goal Orientation', *Psychoneuroendocrinology 44,* 60–70.

Alred, D. (2016), *The Pressure Principle*, London: Penguin.

Amita, S., Prabhakar, S., Manoj, I., Harminder, S., & Pavan, T.
(2009), 'Effect of yoga-nidra on blood glucose level in diabetic
patients', *Indian Journal of Physiology and Pharmacology,*
53 (1), 97–101.

Borchardt, A. R., Patterson, S. M., & Seng, E. K. (2012), 'The effect of
meditation on cortisol: A comparison of meditation techniques
to a control group', Ohio University: Department of Experimental
Health Psychology. Retrieved from http://www.irest.us/sites/
default/files/Meditation%20on%20Cortisol%2012.pdf

Crocker, J., Olivier, M., & Nuer, N. (2009), 'Self-image Goals and
Compassionate Goals: Costs and Benefits', *Self and Identity, 8,*
251–69.

Feldman Barrett, L. (2017), *How Emotions Are Made. The Secret
Life of The Brain*, London: Pan Macmillan.

References

Frederickson, L. B. (2003), 'The Value of Positive Emotions', *American Scientist*, USA: Sigma.

Huberman (2021). Talks by Professor Andrew Huberman on his podcast The Huberman Lab can be accessed on YouTube.

Inagaki, T. K., & Eisenberger, N. I. (2012), 'Neural Correlates of Giving Support to a Loved One', *Psychosomatic Medicine, 74*, 3–7.

Jamieson, J. P., Crum, A.J., Goyer, P., Marotta, M. E., & Akinola, M. (2018), 'Optimizing stress responses with reappraisal and mindset interventions: an integrated model', *Stress, Anxiety & Coping: An International Journal, 31*, 245–61.

Kristensen, T. S., Biarritz, M., Villadsen, E., & Christensen, K. B. (2005), 'The Copenhagen Burnout Inventory: A new tool for the assessment of burnout', *Work & Stress, 19* (3), 192–207.

Kumari, M., Shipley, M., Stafford, M., & Kivimaki, M. (2011), 'Association of diurnal patterns in salivary cortisol with all-cause and cardiovascular mortality: findings from the Whitehall II Study', *Journal of Clinical Endocrinology and Metabolism, 96* (5), 1478–85.

Maslach, C., Jackson, S. E., & Leiter, M. P (1996), *Maslach Burnout Inventory* (3rd ed), Palo Alto, CA: Consulting Psychologists Press.

McEwen, B. S., & Gianaros, P. J. (2010), 'Stress- and Allostasis-induced Brain Plasticity', *Annual Review of Medicine, 62*, 431–45.

McEwen, B. S. (2000), 'The Neurobiology of Stress: from serendipity to clinical relevance', *Brain Research, 886*, 172–89.

McGonigal, K. (2012), *The Willpower Instinct*, London: Avery.

Mogilner, C., Chance, Z., & Norton, M. I. (2012), 'Giving Time Gives You Time', *Psychological Science, 23* (10), 1233–8.

Moszeik, E. N., von Oertzen, T., & Renner, K. H., 'Effectiveness of a short Yoga Nidra meditation on stress, sleep, and well-being in a large and diverse sample', *Current Psychology* (2020), https://doi.org/10.1007/s12144-020-01042-2

Osmo, F., Duran, V., Wenzel, A., et al. (2018), 'The Negative Core Beliefs Inventory (NCBI): Development and Psychometric Properties', *Journal of Cognitive Psychotherapy, 32* (1), 1–18.

Sapolsky, R. (2017), *Behave. The Biology of Humans at Our Best and Worst*, London: Vintage.

Stellar, J. E., John-Henderson, N., Anderson, C. L., Gordon, A. M., McNeil, G. D., & Keltner, D. (2015), 'Positive affect and markers of inflammation: discrete positive emotions predict lower levels of inflammatory cytokines', *Emotion 15* (2), 129–33.

Strack, J., & Esteves, F. (2014), 'Exams? Why Worry? The Relationship Between Interpreting Anxiety as Facilitative, Stress Appraisals, Emotional Exhaustion, and Academic Performance', *Anxiety, Stress, and Coping: An International Journal*, 1–10.

Ware, B. (2012), *The Top Five Regrets of the Dying*, London: Hay House.

Section 8: On a Meaningful Life

Clear, J., *Atomic Habits* (2018), London: Random House.

Feldman Barrett, L. (2017), *How Emotions Are Made. The Secret Life of The Brain*, London: Pan Macmillan.

Fletcher, E. (2019), *Stress Less, Accomplish More*, London: William Morrow.

Gottman, J. M., & Silver, N. (1999), *The Seven Principles for Making Marriage Work*, London: Orion.

References

Hari, J. (2018), *Lost Connections*, London: Bloomsbury.

Johnson, S. (2008), *Hold Me Tight*. London: Piatkus.

Sapolsky, R. (2017), *Behave. The Biology of Humans at Our Best and Worst*, London: Vintage.

Siegel, D. J., & Hartzell, M. (2004), *Parenting from the Inside Out: How a deeper self-understanding can help you raise children who thrive*, New York: Tarcher Perigee.

Thomas, M. (2021), *The Lasting Connection*, London: Robinson.

Waldinger, R. (2015), *What makes a good life? Lessons from the longest study on happiness*, TEDx Beacon Street. https://www.ted.com/talks/robert_waldinger_what_makes_a_good_life_lessons_from_the_longest_st udy_on_happiness/transcript?rid=J7CiE5vP5l5t

Ware, B. (2012), *The Top Five Regrets of the Dying*, London: Hay House.

Diagrams

Figure 1 is an adapted variation based on an original from: Clarke, I., & Wilson, H. (2009), *Cognitive Behaviour Therapy for Acute Inpatient Mental Health Units: Working with Clients, Staff and the Milieu*, Abingdon: Routledge.

Figure 2 is an adapated variation based on an original from: Greenberger, D., & Padesky, C. A. (2016), *Mind Over Mood*, 2nd Edition, New York: Guilford Press.

Figure 3 is an adapted variation based on an original from: Clarke, I., & Wilson, H. (2009), Cognitive Behavioural Therapy for Acute Inpatient Mental Health Units, Sussex: Routledge.

Resources

This book is your toolkit to use for improving or enhancing your mental health and wellbeing. If you find a specific tool or approach especially helpful and are interested to find out more about that approach, see the following list of related self-help books and organizations that offer support.

Isabel Clarke, *How to Deal with Anger: A 5-step CBT-based Plan for Managing Anger and Frustration*, London: Hodder & Stoughton, 2016.

Paul Gilbert, *Overcoming Depression: A self-help guide using Cognitive Behavioural Techniques*, London: Robinson, 1997.

John Gottman & Nan Silver, *The Seven Principles for Making Marriage Work*, London: Orion, 1999.

Alex Korb, *The Upward Spiral: Using neuroscience to reverse the course of depression, one small change at a time*, Oakland, CA: New Harbinger, 2015.

Professor Felice Jacka, *Brain Changer: How diet can save your mental health*, London: Yellow Kite, 2019.

Dr Sue Johnson, *Hold Me Tight*, London: Piatkus, 2008.

Helen Kennerley, *Overcoming Anxiety: A Self-Help Guide Using Cognitive Behavioural Techniques*, London: Robinson, 2014.

Resources

Kristin Neff & Christopher Germer, *The Mindful Self-Compassion Workbook*, New York: Guilford Press, 2018.

Joe Oliver, Jon Hill & Eric Morris, *ACTivate Your Life: Using Acceptance and Mindfulness to Build a Life that is RIch, Fulfilling and Fun*, London: Robinson, 2015.

Julia Samuel, *Grief Works*, London: Penguin Life, 2017.

Michaela Thomas, *The Lasting Connection: Developing Love and Compassion for Yourself and Your Partner*, London: Robinson, 2021.

Organizations that offer support

NHS Choices (UK) – www.nhs.uk

Mind – A charity that offers information on their website and local support initiatives. See www.mind.org.uk

Young Minds – A charity that provides information for children, young people and their parents. See www.youngminds.org.uk

Nightline Association – A service run by students for students through universities. They offer a free, confidential listening service and information. See www.nightline.ac.uk

Samaritans – For anyone in crisis, this service offers support and advice 24 hours a day, 7 days a week. See www.samaritans.org

Acknowledgements

Lots of brilliant individuals have played a part in bringing this book to life. None more than my husband, Matthew. Thank you for stepping up to every single role that this wild journey has demanded. You have been researcher, creative director, videographer, ideas man, editor, business partner, advisor, home schooler, friend, fan, critic and everything in between. You have believed in me relentlessly, even when I didn't believe in myself.

Thank you to my beautiful babies, Sienna, Luke and Leon, for your patience. I have missed you intensely while writing. I hope being a part of this has inspired you to reach for your dreams too. Book or no book, you each remain my greatest achievements and fill me with more pride than any job ever could.

Thank you to my parents who went above and beyond as always to give my children the best home from home when I needed to write. Everything I ever achieve is because you both worked your guts off to provide me with the opportunities that you never had. For that I am grateful every day. Thank you to Pat and David for always being so supportive and encouraging.

Thank you to Francesca Scambler for making the call and taking a chance on me. Thank you to Abigail Bergstrom, my literary agent, who inspired me from the very beginning. I am privileged to have worked with you.

Acknowledgements

A special thank you to my manager, Zara Murdoch. You have been an incredible guide, mentor and all-round superhero. And to Grace Nicholson who completes the dream team and helps to make all this possible.

Thank you to my editor, Ione Walder, for your patience and kindness in helping me turn this text into a book I could be proud of. Thank you to Daniel Bunyard for seeing something in my proposal and inspiring me to work with Penguin on this. Thank you also to Ellie Hughes, Clare Parker, Lucy Hall, Vicky Photiou, Paula Flanagan, Aggie Russell, Lee Motley, Beth O'Rafferty, Nick Lowndes, Emma Henderson and Jane Kirby for all the work behind the scenes at Penguin.

Thank you to Amanda Hardy and Jessica Mason for cheer-leading from the very beginning and listening without judgement when I needed to moan about how tiring this incredible opportunity has been. Thank you to Jackie for looking me straight in the eye and telling me that I could do this in a moment when I needed to hear it, and for making sure I could have it all without having to do it all.

Thank you to all my clients over the years. I have learned more from each of you than I have taught in return and I feel privileged to have walked alongside you on your journeys.

And thank you to every single person who hit follow on any of my social media accounts. We have built such a kind and inspired community. I hope this book helps you to face life armed with some more of the tools you need.

I must give credit to the incredible minds that have worked hard to develop evidence-based psychological therapies and whose work so many people have benefitted from. Please accept my apologies if there are any errors or omissions in how I have translated it.

My Instagram @DrJulie contains videos I have made on the subjects in this book.

Index

acceptance 164, 179–83, 311–13
of death 135, 138, 226–8
Acceptance and Commitment
Therapy (ACT) 225, 228,
287–8
addiction 78, 88, 157,
197, 247
adrenaline 192, 202, 236, 240,
244–5
affirmations 168–9, 266–7
afterlife 228
alcohol 20–21, 115, 129,
175, 196
all-or-nothing thinking
26–8, 31
Alred, Dave 267
anger 133–4, 136, 156, 182–3
anhedonia 68–70
anticipation 89, 151, 197, 236
antidepressants 55
anxiety 115, 124, 154,
189–98, 254
calming tips 59, 200–203

dealing with anxious
thoughts 205–21
and stress 77, 106, 233–7
things that make it worse
195–8
appreciation, expressing 308
arrogance 167
attachment 302–7, 310–11
anxious 303–4
avoidant 304–5
disorganized 306–7
secure 305–6
attention, spotlight of 153–4,
213–15, 266–7, 269
autopilot 29, 38, 83, 107
avoidance 196–7
avoidant attachment 304–5
awe 260–62

bad days 44–51
bargaining 134
baths 58, 110
beliefs 24–5, 195, 228–9, 272

black-and-white thinking
    26–8, 31
blame 310
boundaries 121, 148, 153, 302
breathing
    calming techniques 200–202,
        251–2
    controlling 192
burnout 86, 245–6, 254

caffeine 59
calming 110–12, 200–203, 251–2
cancer 225
caring for others 120–24, 253
Carrey, Jim 34, 175
catastrophizing 209, 216, 266
change 91–6
coaching 169, 177, 218, 266
coffee 42, 53, 61, 100
Cognitive Analytic Therapy
    (CAT) 298
Cognitive Behavioural Therapy
    (CBT) 56, 102
compassion 122, 218, 308
    towards self 79–80, 166,
        179–80, 185–6
Compassion Focused Therapy
    (CFT) 171, 182
confidence, building 162–71

connection 308–9
cortisol 235–6
couples therapy 298, 310
crisis plans 88, 122
criticism 311–12
    dealing with 151–60, 169,
        173–4
    self-criticism 35, 46–7,
        78–80, 158, 169, 184–5,
        271–2
curiosity 102

dancing 56
death 127–36
    accepting 226–9
    beliefs about 228–9
    fear of 223–30, 228–9
    see also grief
decision-making 44–5, 78,
    196, 244
denial 132–3
depression 69–70, 114–15, 254
    and exercise 55–6
    and grief 129, 134–5
    see also low mood
Dialectical Behaviour Therapy
    (DBT) 90–91, 110–11
diet 53, 59–61, 108, 147
disapproval, dealing with 151–60

dopamine 55, 87
drugs 20

egocentric thinking 23–4, 27,
    155–6
Einstein, Albert 94
emotional reasoning 24, 27
emotions
  being responsive 182,
    310–11
  expressing 141–2, 147
  happiness 277–82
  naming and describing
    108–9, 114–18, 182–3
  painful 99–118, 129, 133–4,
    139–40
  understanding 16–18
encouragement 47, 79, 169
enthusiasm 72, 87, 279, 285
escaping 120, 196, 206, 228,
    235, 255
exams 24, 26, 47, 95, 219–20,
    234, 264
exercise 13–14, 53, 55, 58, 73,
    147, 202–3

fact checking 212–13
failure 78–80, 173–7, 269–72
fame 224

fear 132–3, 164, 170
  acknowledging 192–3
  dealing with 189–98
  of death 223–30
feedback 160, 311
Feeling Wheel 117
fight or flight response 206,
    235, 236, 253
focus 268–9
food 20, 59–60, 77
friendship 40, 110, 112, 297
  strengthening 307–8
future 90–91

Gilbert, Paul 171
Gilovich, Thomas 153
goals 280–81
  setting 46, 91–2, 254–5,
    293–4
  shared 314
  small and specific 87, 92
  staying connected with 74
Gottman, John 307
gratitude 42, 88, 261, 313
grief 127–36, 227
  counselling for 144
  healing and rebuilding
    143–4, 146–8
  mourning 138–44

Index

grief – (*cont.*)
  stages of 132–6
  unresolved 129
grit 86
grounding techniques 139
guilt 140, 156, 271

habits, developing 76, 84–5
happiness 9, 114, 254,
  277–82, 297
The Happiness Trap 278
Harris, Russ 278
health, looking after 45
health anxiety 224
heart rate 67, 77, 100
heights, fear of 189–92, 224
help, seeking 317–19
hospitals 217

identity 220–21
  and goals 89–90, 92
immune system 244–5
insomnia 12, 53, 57–9, 61

Johnson, Sue 310
journalling 29, 74, 92, 95–6,
  118, 180
  about death 229–30
  about relationships 314–15

kindness 80, 123, 217–18,
  257, 285
Kubler-Ross, Elisabeth 132

labelling 211–12
language 266–7, 273
  and emotions 114–18
laziness 179
Learning Zone Model
  164–5
Lee, Deborah 171
light, exposure to 58
listening 122–3
loneliness 62, 254
low mood 9–19
  downward spiral 14–16
  formulations for 48–9
  instant relief, vicious cycle of
    20–21
  masking 9–11
  things that make it
    worse 22
  see also depression

The Mask 175, 334
meaningful life 293–5
meditation 255–6
memory 100, 139, 150
mental filters 24–5, 210–11

mental health
  seeking therapy 317–19
  working on 4–5
metacognition 35, 94–5
mind reading 22–3, 27
mindfulness 30, 36–8, 41, 84,
  105–7, 207–8
  in everyday life 257–60
mindsets 13, 265–6
mistakes 157, 173–7
mood see low mood
motivation 67–70, 82–92
  nurturing 72–80
mourning 138–44
music 56, 73, 110, 112, 116, 142
'musts' 25–6, 27

negative thoughts 35, 210
noradrenaline 87
nutrition 59–61

opposite action skill 83–4
outdoors 55, 58, 87, 262
overgeneralization 23, 27, 211

panic attacks 100, 224
parenting 245, 253, 278
  and attachment 302, 305
parties 170

people-pleasing 152–4, 303
perfectionism 26, 44
perseverance 86
personalizing 209–10
phobias 189–92, 224, 226
phones 86, 198
Pisa, Leaning Tower of
  189–92, 210
planning 88–9
prejudice 167
problem solving 94–5
procrastination 68–70
progress, acknowledging 75

reassurance, seeking 197
reflecting 95–6
reframing 219–20, 268, 270
rejection 115
relationships 62–4, 253,
  297–315
  conflict 311–12
  disagreements 299
  ending 301
  healthy 298, 308–15
  myths 299–301
  repairing 312–13
  see also friendship
relaxation 134, 252
resentment 180

Index

resilience 157, 175, 253, 272–3
rest, best types of 86–7
rewards 85, 87
risk-taking 174, 223–4, 226
routine 61–2, 148
rumination 39–40
running 106, 293–4

safety behaviours 198
Samuel, Julia 146
Savitsky, Kenneth 153
screen time 58, 86
secure attachment 305–6
self-acceptance 179–86
self-awareness 17, 148, 180,
    309–10
self-control 76–8
self-criticism 35, 46–7, 78–80,
    158, 169, 184–5, 271–2
self-doubt 1, 151–60, 173–82
self-esteem 166–7
self-soothing 109–12, 139,
    274, 304
sensitization 197
sexual problems 247
shame 78–9, 156–9, 174, 311
    resilience against 272–3
shock 132
'shoulds' 25–6, 27

showering 259
sleep 11–12, 57–9, 61, 78,
    108, 247
smells, calming 111–12
social isolation 63, 253
social media 20, 25, 173–4,
    210, 277
solutions, focusing on 50–51
spotlight effect 153–4, 213–15,
    266–7, 269
square breathing 201–2
stress
    anticipatory 236
    and anxiety 233–7
    chronic 245–6
    importance of 239–42
    managing 106, 251–62,
        264–74
    negative impact 244–9
    reducing 239–42
suicide 129
supporting others 120–24
Sydney Harbour Bridge 213–15
symptoms, physical 100

talking 141–2
tea 112
television 20, 61
temptation, resisting 76–7

therapy, when to seek 317–19
thought biases 21–34, 39,
   208–13, 272
thoughts
   anxious 205–21
   distancing from 207–8
   negative 13–14, 102
   reframing 219–20
threat response 110, 156
tunnel vision 268–9

values 177, 220–21, 246, 255,
   279–82
   clarifying and adjusting
   284–91, 294

and a meaningful life
   293–5
shared 313–14
vocabulary 114–18
vulnerability 165, 306–7

walking, mindful 258
willpower 76–8
Worden, William 138
worry lists 58–9

Yalom, Irvin 224
yoga 56
   nidra 256
YouTube 2, 41

Spare tools

Here are some blank reproductions of the tools found throughout the book for you to have a go at filling out yourself.

The cross-sectional formulation

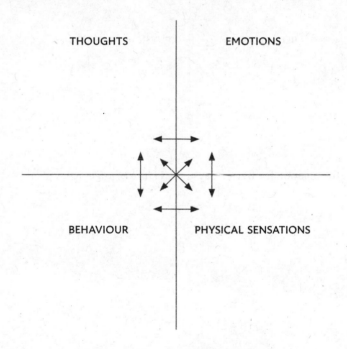

Blank formulation for low mood (see Figure 5, page 48).

The cross-sectional formulation

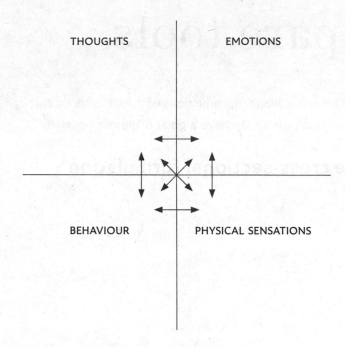

Blank formulation for better days (see Figure 6, page 50).

The values chart

Use these blank grids to help you reflect on what you value most in each area of your life (see page 286).

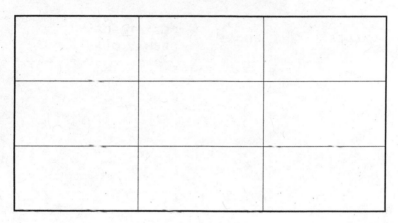

Values, goals, actions

Use these blank charts to help you translate your values into goals and everyday actions (see Figure 11, page 285).

VALUES	GOALS	DAY-TO-DAY BEHAVIOURS

VALUES	GOALS	DAY-TO-DAY BEHAVIOURS

The values star

Here are a few more blank values stars for you to fill in using Figure 12 on page 289 to help you.

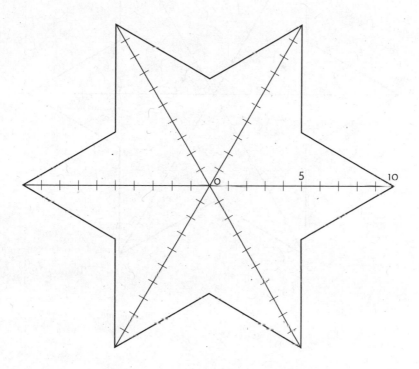

Spare tools

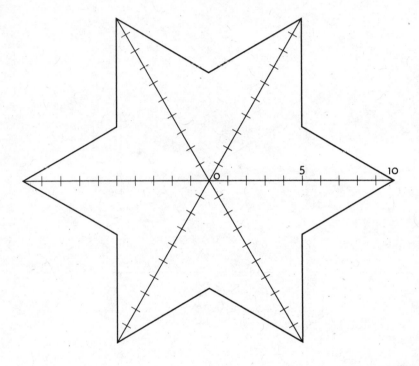